IT'S HERE!

PRENTICE HALL SCIENCE

FINALLY, THE PERFECT FIT.

NOW YOU CAN CHOOSE THE PERFECT FIT FOR ALL YOUR CURRICULUM NEEDS.

The new Prentice Hall Science program consists of 19 hardcover books, each of which covers a particular area of science. All of the sciences are represented in the program so you can choose the perfect fit to *your* particular curriculum needs.

The flexibility of this program will allow you to teach those topics you want to teach, and to teach them *in-depth*. Virtually any approach to science—general, integrated, coordinated, thematic, etc.—is possible with Prentice Hall Science.

Above all, the program is designed to make your teaching experience easier and more fun.

ELECTRICITY AND MAGNETISM
Ch. 1. Electric Charges and Currents
Ch. 2. Magnetism
Ch. 3. Electromagnetism
Ch. 4. Electronics and Computers

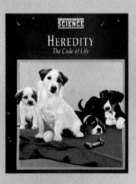

HEREDITY: THE CODE OF LIFE
Ch. 1. What is Genetics?
Ch. 2. How Chromosomes Work
Ch. 3. Human Genetics
Ch. 4. Applied Genetics

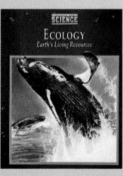

ECOLOGY: EARTH'S LIVING RESOURCES
Ch. 1. Interactions Among Living Things
Ch. 2. Cycles in Nature
Ch. 3. Exploring Earth's Biomes
Ch. 4. Wildlife Conservation

PARADE OF LIFE: MONERANS, PROTISTS, FUNGI, AND PLANTS
Ch. 1. Classification of Living Things
Ch. 2. Viruses and Monerans
Ch. 3. Protists
Ch. 4. Fungi
Ch. 5. Plants Without Seeds
Ch. 6. Plants With Seeds

EXPLORING THE UNIVERSE
Ch. 1. Stars and Galaxies
Ch. 2. The Solar System
Ch. 3. Earth and Its Moon

EVOLUTION: CHANGE OVER TIME
Ch. 1. Earth's History in Fossils
Ch. 2. Changes in Living Things Over Time
Ch. 3. The Path to Modern Humans

EXPLORING EARTH'S WEATHER
Ch. 1. What Is Weather?
Ch. 2. What Is Climate?
Ch. 3. Climate in the United States

THE NATURE OF SCIENCE
Ch. 1. What is Science?
Ch. 2. Measurement and the Sciences
Ch. 3. Tools and the Sciences

ECOLOGY:
EARTH'S NATURAL RESOURCES

Ch. 1. Energy Resources
Ch. 2. Earth's Nonliving
Resources
Ch. 3. Pollution
Ch. 4. Conserving Earth's
Resources

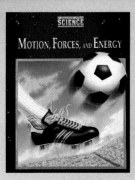

MOTION, FORCES,
AND ENERGY

Ch. 1. What Is Motion?
Ch. 2. The Nature of Forces
Ch. 3. Forces in Fluids
Ch. 4. Work, Power, and
Simple Machines
Ch. 5. Energy: Forms
and Changes

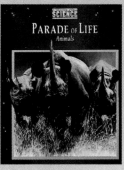

PARADE OF LIFE: ANIMALS

Ch. 1. Sponges, Cnidarians,
Worms, and Mollusks
Ch. 2. Arthropods and
Echinoderms
Ch. 3. Fish and Amphibians
Ch. 4. Reptiles and Birds
Ch. 5. Mammals

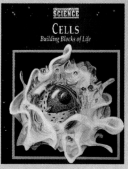

CELLS:
BUILDING BLOCKS OF LIFE

Ch. 1. The Nature of LIfe
Ch. 2. Cell Structure and
Function
Ch. 3. Cell Processes
Ch. 4. Cell Energy

DYNAMIC EARTH

Ch. 1. Movement of the
Earth's Crust
Ch. 2. Earthquakes and
Volcanoes
Ch. 3. Plate Tectonics
Ch. 4. Rocks and Minerals
Ch. 5. Weathering and
Soil Formation
Ch. 6. Erosion and Deposition

MATTER: BUILDING BLOCK OF
THE UNIVERSE

Ch. 1. General Properties
of Matter
Ch. 2. Physical and Chemical
Changes
Ch. 3. Mixtures, Elements, and
Compounds
Ch. 4. Atoms: Building Blocks
of Matter
Ch. 5. Classification of Elements:
The Periodic Table

CHEMISTRY OF MATTER

Ch. 1. Atoms and Bonding
Ch. 2. Chemical Reactions
Ch. 3. Families of Chemical
Compounds
Ch. 4. Chemical Technology
Ch. 5. Radioactive Elements

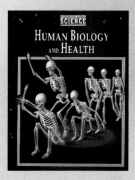

HUMAN BIOLOGY AND
HEALTH

Ch. 1. The Human Body
Ch. 2. Skeletal and Muscular
Systems
Ch. 3. Digestive System
Ch. 4. Circulatory System
Ch. 5. Respiratory and
Excretory Systems
Ch. 6. Nervous and Endocrine
Systems
Ch. 7. Reproduction and
Development
Ch. 8. Immune System
Ch. 9. Alcohol, Tobacco,
and Drugs

EXPLORING PLANET EARTH

Ch. 1. Earth's Atmosphere
Ch. 2. Earth's Oceans
Ch. 3. Earth's Fresh Water
Ch. 4. Earth's Landmasses
Ch. 5. Earth's Interior

HEAT ENERGY

Ch. 1. What Is Heat?
Ch. 2. Uses of Heat

SOUND AND LIGHT

Ch. 1. Characteristics of Waves
Ch. 2. Sound and Its Uses
Ch. 3. Light and the Electro-
magnetic Spectrum
Ch. 4. Light and Its Uses

A COMPLETELY INTEGRATED LEARNING SYSTEM...

The Prentice Hall Science program is an *integrated* learning system with a variety of print materials and multimedia components. All are designed to meet the needs of diverse learning styles and your technology needs.

THE STUDENT BOOK

Each book is a model of **excellent writing and dynamic visuals**—designed to be exciting and motivating to the student *and* the teacher, with relevant examples integrated throughout, and more opportunities for many different activities which apply to everyday life.

Problem-solving activities emphasize the thinking process, so problems may be more open-ended.

"Discovery Activities" throughout the book foster active learning.

Different sciences, and other disciplines, are integrated throughout the text and reinforced in the "Connections" features (the connections between computers and viruses is one example).

TEACHER'S RESOURCE PACKAGE

In addition to the student book, the complete teaching package contains:

ANNOTATED TEACHER'S EDITION

Designed to provide **"teacher-friendly"** support regardless of instructional approach:

■ **Help is readily available** if you choose to teach thematically, to integrate the sciences, and/or to integrate the sciences with other curriculum areas.

■ **Activity-based learning** is easy to implement through the use of Discovery Strategies, Activity Suggestions, and Teacher Demonstrations.

■ Integration of all components is part of the teaching strategies.

■ For instant accessibility, all of the teaching suggestions are wrapped around the student pages to which they refer.

ACTIVITY BOOK

Includes a **discovery activity for each chapter**, plus other activities including problem-solving and cooperative-learning activities.

THE REVIEW AND REINFORCEMENT GUIDE

Addresses **students' different learning styles** in a clear and comprehensive format:

■ Highly visual for visual learners.

TEACHER'S RESOURCE PACKAGE

FOR THE PERFECT FIT TO YOUR TEACHING NEEDS.

■ Can be used in conjunction with the program's audiotapes for auditory and language learners.

■ More than a study guide, it's a guide to comprehension, with activities, key concepts, and vocabulary.

ENGLISH AND SPANISH AUDIOTAPES
Correlate with the Review and Reinforcement Guide to aid auditory learners.

LABORATORY MANUAL ANNOTATED TEACHER'S EDITION
Offers **at least one additional hands-on opportunity per chapter** with

answers and teaching suggestions on lab preparation and safety.

TEST BOOK
Contains **traditional and up-to-the-minute strategies for student assessment.** Choose from performance-based tests in addition to traditional chapter tests and computer test bank questions.

STUDENT LABORATORY MANUAL
Each of the 19 books also comes with its own Student Lab Manual.

ALSO INCLUDED IN THE INTEGRATED LEARNING SYSTEM:

■ Teacher's Desk Reference
■ English Guide for Language Learners
■ Spanish Guide for Language Learners
■ Product Testing Activities
■ Transparencies

■ Computer Test Bank (IBM, Apple, or MAC)
■ VHS Videos
■ Videodiscs
■ Interactive Videodiscs (Level III)
■ Interactive Videodiscs/ CD ROM
■ Courseware

All components are integrated in the teaching strategies in the Annotated Teacher's Edition, where they directly relate to the science content.

THE PRENTICE HALL SCIENCE
INTEGRATED LEARNING SYSTEM

The following components are integrated in the teaching strategies for
PARADE OF LIFE: ANIMALS.

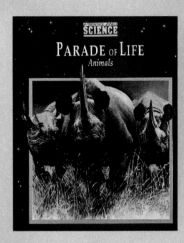

- **Spanish Audiotape English Audiotape**
- **Activity Book**
- **Review and Reinforcement Guide**
- **Test Book**—including Performance-Based Tests
- **Laboratory Manual, Annotated Teacher's Edition**

- **Laboratory Manual**
- **English Guide for Language Learners**
- **Spanish Guide for Language Learners**
- **Videos/Videodiscs:**
  Super Scents
  Seeing Sense
  Sound Sense

- **Interactive Videodiscs:**
  Insects: Little Giants of the Earth
  On Dry Land: The Desert Biome
  ScienceVision: EcoVision
  In the Company of Whales
- **Interactive Videodiscs/ CD ROM:**
  Paul ParkRanger and the Mystery of the Disappearing Ducks
  Amazonia
  Virtual BioPark

INTEGRATING OTHER SCIENCES

Many of the other 18 Prentice Hall Science books can be integrated into **PARADE OF LIFE: ANIMALS.** The books you will find suggested most often in the Annotated Teacher's Edition are PARADE OF LIFE: MONERANS, PROTISTS, FUNGI, AND PLANTS; ECOLOGY: EARTH'S LIVING RESOURCES; CELLS: BUILDING BLOCKS OF LIFE; EVOLUTION: CHANGE OVER TIME; EXPLORING PLANET EARTH; HUMAN BIOLOGY AND HEALTH; SOUND AND LIGHT; and ELECTRICITY AND MAGNETISM.

INTEGRATING THEMES

Many themes can be integrated into **PARADE OF LIFE: ANIMALS.**
Following are the ones most commonly suggested in the Annotated Teacher's Edition: EVOLUTION, PATTERNS OF CHANGE, SCALE AND STRUCTURE, and UNITY AND DIVERSITY.

For more detailed information on teaching thematically and integrating the sciences, see the Teacher's Desk Reference and teaching strategies throughout the Annotated Teacher's Edition.

For more information, call 1-800-848-9500 or write:

 P R E N T I C E   H A L L

Simon & Schuster Education Group
113 Sylvan Avenue Route 9W
Englewood Cliffs, New Jersey 07632
Simon & Schuster A Paramount Communications Company

Annotated Teacher's Edition

Prentice Hall Science

Parade of Life
Animals

Anthea Maton
Former NSTA National
    Coordinator
Project Scope, Sequence,
    Coordination
Washington, DC

Jean Hopkins
Science Instructor and Department
    Chairperson
John H. Wood Middle School
San Antonio, Texas

Susan Johnson
Professor of Biology
Ball State University
Muncie, Indiana

David LaHart
Senior Instructor
Florida Solar Energy Center
Cape Canaveral, Florida

Maryanna Quon Warner
Science Instructor
Del Dios Middle School
Escondido, California

Jill D. Wright
Professor of Science Education
Director of International Field
    Programs
University of Pittsburgh
Pittsburgh, Pennsylvania

Prentice Hall
A Division of Simon & Schuster
Englewood Cliffs, New Jersey

ISBN 0-13-400441-8

 5 6 7 8 9 10   97 96 95

Contents of Annotated Teacher's Edition

To the Teacher   T–3

About the Teacher's Desk Reference   T–3

Integrating the Sciences   T–4

Thematic Overview   T–4

Thematic Matrices   T–5

Comprehensive List of Laboratory Materials   T–10

To the Teacher

Welcome to the *Prentice Hall Science* program. *Prentice Hall Science* has been designed as a complete program for use with middle school or junior high school science students. The program covers all relevant areas of science and has been developed with the flexibility to meet virtually all your curriculum needs. In addition, the program has been designed to better enable you—the classroom teacher—to integrate various disciplines of science into your daily lessons, as well as to enhance the thematic teaching of science.

The *Prentice Hall Science* program consists of nineteen books, each of which covers a particular topic area. The nineteen books in the *Prentice Hall Science* program are

The Nature of Science
Parade of Life: Monerans, Protists, Fungi, and Plants
Parade of Life: Animals
Cells: Building Blocks of Life
Heredity: The Code of Life
Evolution: Change Over Time

Ecology: Earth's Living Resources
Human Biology and Health
Exploring Planet Earth
Dynamic Earth
Exploring Earth's Weather
Ecology: Earth's Natural Resources
Exploring the Universe
Matter: Building Block of the Universe
Chemistry of Matter
Electricity and Magnetism
Heat Energy
Sound and Light
Motion, Forces, and Energy

Each of the student editions listed above also comes with a complete set of teaching materials and student ancillary materials. Furthermore, videos, interactive videos and science courseware are available for the *Prentice Hall Science* program. This combination of student texts and ancillaries, teacher materials, and multimedia products makes up your complete *Prentice Hall Science* Learning System.

About the Teacher's Desk Reference

The *Teacher's Desk Reference* provides you, the teacher, with an insight into the workings of the *Prentice Hall Science* program. The *Teacher's Desk Reference* accomplishes this task by including all the standard information you need to know about *Prentice Hall Science*.

The *Teacher's Desk Reference* presents an overview of the program, including a full description of each ancillary available in the program. It gives a brief summary of each of the student textbooks available in the *Prentice Hall Science* Learning System. The *Teacher's Desk Reference* also demonstrates how the seven science themes incorporated into *Prentice Hall Science* are woven throughout the entire program.

In addition, the *Teacher's Desk Reference* presents a detailed discussion of the features of the Student Edition and the features of the Annotated Teacher's Edition, as well as an overview section that summarizes issues in science education and offers a message about teaching special students. Selected instructional essays in the *Teacher's Desk Reference* include English as a Second Language (ESL), Multicultural Teaching, Cooperative-Learning Strategies, and Integrated Science Teaching, in addition to other relevant topics. Further, a discussion of the Multimedia components that are part of *Prentice Hall Science*, as well as how they can be integrated with the textbooks, is included in the *Teacher's Desk Reference*.

The *Teacher's Desk Reference* also contains in blackline master form a booklet on Teaching Graphing Skills, which may be reproduced for student use.

Integrating the Sciences

The *Prentice Hall Science* Learning System has been designed to allow you to teach science from an integrated point of view. Great care has been taken to integrate other science disciplines, where appropriate, into the chapter content and visuals. In addition, the integration of other disciplines such as social studies and literature has been incorporated into each textbook.

On the reduced student pages throughout your Annotated Teacher's Edition you will find numbers within blue bullets beside selected passages and visuals. An Annotation Key in the wraparound margins indicates the particular branch of science or other discipline that has been integrated into the student text. In addition, where appropriate, the name of the textbook and the chapter number in which the particular topic is discussed in greater detail is provided. This enables you to further integrate a particular science topic by using the complete *Prentice Hall Science* Learning System.

Thematic Overview

When teaching any science topic, you may want to focus your lessons around the underlying themes that pertain to all areas of science. These underlying themes are the framework from which all science can be constructed and taught. The seven underlying themes incorporated into *Prentice Hall Science* are

Energy
Evolution
Patterns of Change
Scale and Structure
Systems and Interactions
Unity and Diversity
Stability

The primary themes in this textbook are Evolution, Patterns of Change, Scale and Structure, and Unity and Diversity. Primary themes throughout *Prentice Hall Science* are denoted by an asterisk.

A detailed discussion of each of these themes and how they are incorporated into the *Prentice Hall Science* program are included in your *Teacher's Desk Reference*. In addition, the *Teacher's Desk Reference* includes thematic matrices for the *Prentice Hall Science* program.

A thematic matrix for each chapter in this textbook follows. Each thematic matrix is designed with the list of themes along the left-hand column and in the right-hand column a big idea, or overarching concept statement, as to how that particular theme is taught in the chapter.

CHAPTER 1

Sponges, Cnidarians, Worms, and Mollusks

ENERGY	• Animals are heterotrophs. They obtain energy by eating autotrophs or other heterotrophs.
*EVOLUTION	• All invertebrates share an evolutionary heritage.
*PATTERNS OF CHANGE	• The levels of organization become higher as animals become more complex.
*SCALE AND STRUCTURE	• The organs and other body structures that carry out an animal's life functions vary from phylum to phylum.
SYSTEMS AND INTERACTIONS	• Invertebrates respond to and interact with their environment in ways that help them gather food, protect themselves, and reproduce.
*UNITY AND DIVERSITY	• Different groups of invertebrates carry out the basic life functions in a variety of ways.
STABILITY	• Although the levels of organization become higher as animals become more complex, all animals maintain a stable internal environment.

CHAPTER 2

Arthropods and Echinoderms

ENERGY	• Arthropods and echinoderms obtain energy by eating plants or other animals.
*EVOLUTION	• Arthropods and echinoderms have evolved from a common ancestor.
*PATTERNS OF CHANGE	• The levels of organization become higher as animals become more complex.
*SCALE AND STRUCTURE	• The life functions of arthropods and echinoderms vary from phylum to phylum along with their body structures.
SYSTEMS AND INTERACTIONS	• Arthropods and echinoderms respond to and interact with their environment in ways that help them gather food, reproduce, and protect themselves.
*UNITY AND DIVERSITY	• Arthropods and echinoderms perform the basic life functions in different ways.
STABILITY	• The various life functions of arthropods and echinoderms work together to maintain a stable internal and external environment.

CHAPTER 3

Fishes and Amphibians

ENERGY	• Fishes and amphibians obtain energy by eating plants or other animals.
*EVOLUTION	• Fishes were the first vertebrates to have evolved. Amphibians are thought to have evolved from early lobe-finned, bony fishes with lungs.
*PATTERNS OF CHANGE	• Many amphibians undergo a series of dramatic changes in body form, which is known as metamorphosis.
*SCALE AND STRUCTURE	• The body structures of fishes and amphibians vary along with their life functions.
SYSTEMS AND INTERACTIONS	• Fishes and amphibians respond to and interact with their environment in ways that help them gather food, reproduce, and protect themselves.
*UNITY AND DIVERSITY	• Although fishes and amphibians are different in habits and appearances, both groups are chordates.
STABILITY	• The various life functions of fishes and amphibians serve to maintain a stable internal and external environment for these vertebrates.

CHAPTER 4

Reptiles and Birds

ENERGY	• To meet the energy demands of warmbloodedness and flight, birds must acquire a lot of energy in the form of food.
***EVOLUTION**	• Reptiles appeared hundreds of millions of years ago, soon after the first amphibians. • Birds evolved comparatively recently from reptile ancestors, which were probably dinosaurs.
***PATTERNS OF CHANGE**	• Reptiles and birds, unlike amphibians, do not undergo metamorphosis. Development is basically completed while in the egg. • Birds migrate in response to seasonal changes in food supply.
***SCALE AND STRUCTURE**	• The shell and membranes of a reptilian egg make it possible for it to be laid on land. • The shapes and colors of feathers help them to perform their functions.
SYSTEMS AND INTERACTIONS	• The body systems of reptiles make them and their descendants better suited to life on land than amphibians and enable reptiles to inhabit all sorts of land environments. • Bird body systems show many adaptations for flight. • Birds communicate with other members of their species and of their community.
***UNITY AND DIVERSITY**	• Reptiles may be egg-laying or live-bearing. • The amount of parental care among reptiles and birds varies greatly. • Although they vary quite a bit, all vertebrates with feathers are classified as birds.
STABILITY	• Certain reptiles and birds return year after year to the same places to breed and/or lay their eggs. • Feathers help to insulate birds.

CHAPTER 5

Mammals

ENERGY	• Mammals obtain energy by eating plants or other animals.
***EVOLUTION**	• The first mammals, which appeared 200 million years ago, evolved from a now-extinct group of reptiles.
***PATTERNS OF CHANGE**	• The differences in the way mammals reproduce provide a means of classifying them into the following three groups: egg-laying mammals, pouched mammals, and placental mammals.
***SCALE AND STRUCTURE**	• The body structures of mammals vary according to their life functions.
SYSTEMS AND INTERACTIONS	• Mammals respond to and interact with their environment in ways that help them gather food, reproduce, and protect themselves.
***UNITY AND DIVERSITY**	• Although each group of mammals has different structures, they are all chordates.
STABILITY	• The various life functions of mammals serve to maintain a stable internal and external environment.

Comprehensive List of Laboratory Materials

Item	Quantities per Group	Chapter
Aluminum foil	1 sheet	2
Aquarium, rectangular, with cover, filter, and optional light	1	3
Cardboard	1 piece	1
Comb (or brush)	1 per student	5
Coverslip	1 per student	5
Dip net	1	3
Dissecting needle	1	4
Earthworms, live, in a storage container	2	1
Elodea or other aquatic plants	several	3
Glass slide	1 per student	5
Gravel	to fill bottom of aquarium	3
Guppies	several	3
Guppy food	as needed	3
Hand lens	1 per student	5
Isopods	10	2
Jar	1	2
Lamp, desk	1	1
Light, electric	1 per student	5
Magnifying glass	1	4
Medicine dropper	1 per group	1
	1 per student	5
Methylene blue	small bottle	5
Metric ruler	1	3, 4
Microscope	1 per student	5
Owl pellet	1	4
Paper towels	2	1, 2
Scissors	1 pair per student	5
Shoe box with lid	1	2
Snails	several	3
Tape, masking	1 small roll	2
Thermometer	1	3
Tray	1	1

PRENTICE HALL SCIENCE

PARADE OF LIFE
Animals

Anthea Maton
Former NSTA National Coordinator
Project Scope, Sequence, Coordination
Washington, DC

Jean Hopkins
Science Instructor and Department Chairperson
John H. Wood Middle School
San Antonio, Texas

Susan Johnson
Professor of Biology
Ball State University
Muncie, Indiana

David LaHart
Senior Instructor
Florida Solar Energy Center
Cape Canaveral, Florida

Maryanna Quon Warner
Science Instructor
Del Dios Middle School
Escondido, California

Jill D. Wright
Professor of Science Education
Director of International Field Programs
University of Pittsburgh
Pittsburgh, Pennsylvania

Prentice Hall
Englewood Cliffs, New Jersey
Needham, Massachusetts

Prentice Hall Science
Parade of Life: Animals

Student Text and Annotated Teacher's Edition
Laboratory Manual
Teacher's Resource Package
Teacher's Desk Reference
Computer Test Bank
Teaching Transparencies
Product Testing Activities
Computer Courseware
Video and Interactive Video

The illustration on the cover, rendered by Joseph Cellini, depicts a herd of rhinos, which are among the endangered animals that live in Africa.

Credits begin on page 176.

SECOND EDITION

ISBN 0-13-400433-7

5 6 7 8 9 10   97 96 95

 Prentice Hall
A Division of Simon & Schuster
Englewood Cliffs, New Jersey 07632

STAFF CREDITS

Editorial:	Harry Bakalian, Pamela E. Hirschfeld, Maureen Grassi, Robert P. Letendre, Elisa Mui Eiger, Lorraine Smith-Phelan, Christine A. Caputo
Design:	AnnMarie Roselli, Carmela Pereira, Susan Walrath, Leslie Osher, Art Soares
Production:	Suse F. Bell, Joan McCulley, Elizabeth Torjussen, Christina Burghard
Photo Research:	Libby Forsyth, Emily Rose, Martha Conway
Publishing Technology:	Andrew Grey Bommarito, Deborah Jones, Monduane Harris, Michael Colucci, Gregory Myers, Cleasta Wilburn
Marketing:	Andrew Socha, Victoria Willows
Pre-Press Production:	Laura Sanderson, Kathryn Dix, Denise Herckenrath
Manufacturing:	Rhett Conklin, Gertrude Szyferblatt

Consultants

Kathy French	National Science Consultant
Jeannie Dennard	National Science Consultant
Brenda Underwood	National Science Consultant
Janelle Conarton	National Science Consultant

Contributing Writers

Linda Densman
Science Instructor
Hurst, TX

Linda Grant
Former Science Instructor
Weatherford, TX

Heather Hirschfeld
Science Writer
Durham, NC

Marcia Mungenast
Science Writer
Upper Montclair, NJ

Michael Ross
Science Writer
New York City, NY

Content Reviewers

Dan Anthony
Science Mentor
Rialto, CA

John Barrow
Science Instructor
Pomona, CA

Leslie Bettencourt
Science Instructor
Harrisville, RI

Carol Bishop
Science Instructor
Palm Desert, CA

Dan Bohan
Science Instructor
Palm Desert, CA

Steve M. Carlson
Science Instructor
Milwaukie, OR

Larry Flammer
Science Instructor
San Jose, CA

Steve Ferguson
Science Instructor
Lee's Summit, MO

Robin Lee Harris
Freedman
Science Instructor
Fort Bragg, CA

Edith H. Gladden
Former Science Instructor
Philadelphia, PA

Vernita Marie Graves
Science Instructor
Tenafly, NJ

Jack Grube
Science Instructor
San Jose, CA

Emiel Hamberlin
Science Instructor
Chicago, IL

Dwight Kertzman
Science Instructor
Tulsa, OK

Judy Kirschbaum
Science/Computer Instructor
Tenafly, NJ

Kenneth L. Krause
Science Instructor
Milwaukie, OR

Ernest W. Kuehl, Jr.
Science Instructor
Bayside, NY

Mary Grace Lopez
Science Instructor
Corpus Christi, TX

Warren Maggard
Science Instructor
PeWee Valley, KY

Della M. McCaughan
Science Instructor
Biloxi, MS

Stanley J. Mulak
Former Science Instructor
Jensen Beach, FL

Richard Myers
Science Instructor
Portland, OR

Carol Nathanson
Science Mentor
Riverside, CA

Sylvia Neivert
Former Science Instructor
San Diego, CA

Jarvis VNC Pahl
Science Instructor
Rialto, CA

Arlene Sackman
Science Instructor
Tulare, CA

Christine Schumacher
Science Instructor
Pikesville, MD

Suzanne Steinke
Science Instructor
Towson, MD

Len Svinth
Science Instructor/
Chairperson
Petaluma, CA

Elaine M. Tadros
Science Instructor
Palm Desert, CA

Joyce K. Walsh
Science Instructor
Midlothian, VA

Steve Weinberg
Science Instructor
West Hartford, CT

Charlene West, PhD
Director of Curriculum
Rialto, CA

John Westwater
Science Instructor
Medford, MA

Glenna Wilkoff
Science Instructor
Chesterfield, OH

Edee Norman Wiziecki
Science Instructor
Urbana, IL

Teacher Advisory Panel

Beverly Brown
Science Instructor
Livonia, MI

James Burg
Science Instructor
Cincinnati, OH

Karen M. Cannon
Science Instructor
San Diego, CA

John Eby
Science Instructor
Richmond, CA

Elsie M. Jones
Science Instructor
Marietta, GA

Michael Pierre
McKereghan
Science Instructor
Denver, CO

Donald C. Pace, Sr.
Science Instructor
Reisterstown, MD

Carlos Francisco Sainz
Science Instructor
National City, CA

William Reed
Science Instructor
Indianapolis, IN

Multicultural Consultant

Steven J. Rakow
Associate Professor
University of Houston—
Clear Lake
Houston, TX

English as a Second Language (ESL) Consultants

Jaime Morales
Bilingual Coordinator
Huntington Park, CA

Pat Hollis Smith
Former ESL Instructor
Beaumont, TX

Reading Consultant

Larry Swinburne
Director
Swinburne Readability
Laboratory

CONTENTS

PARADE OF LIFE: ANIMALS

CHAPTER 1 Sponges, Cnidarians, Worms, and Mollusks 10

1–1 The Five Kingdoms.....................12
1–2 Introduction to the Animal Kingdom..........15
1–3 Sponges 18
1–4 Cnidarians 20
1–5 Worms 25
1–6 Mollusks 30

CHAPTER 2 Arthropods and Echinoderms 38

2–1 Arthropods: The "Joint-Footed" Animals 40
2–2 Insects: The Most Numerous Arthropods 48
2–3 Echinoderms: The "Spiny-Skinned" Animals 54

CHAPTER 3 Fishes and Amphibians.........................62

3–1 What Is a Vertebrate?.................64
3–2 Fishes 67
3–3 Amphibians.............................76

CHAPTER 4 Reptiles and Birds 88

4–1 Reptiles.................................90
4–2 Birds 103

CHAPTER 5 Mammals 118

5–1 What Is a Mammal?....................120
5–2 Egg-Laying Mammals 123
5–3 Pouched Mammals.....................125
5–4 Placental Mammals 127

SCIENCE GAZETTE

Michael Werikhe: Saving the Rhino—
  One Step at a Time.........................**142**
Tuna Net Fishing: Dolphins in Danger!............**144**
The HIT of the Class of 2025............................**147**

Activity Bank/Reference Section

For Further Reading	**150**
Activity Bank	**151**
Appendix A: The Metric System	**166**
Appendix B: Laboratory Safety: Rules and Symbols	**167**
Appendix C: Science Safety Rules	**168**
Glossary	**170**
Index	**172**

Features

Laboratory Investigations
Observing Earthworm Responses ... 34
Investigating Isopod Environments ... 58
Designing an Aquatic Environment ... 84
Owl Pellets ... 114
Examining Hair ... 138

Activity: Discovering
Observing a Sponge ... 19
Worms at Work ... 29
The Flour Beetle ... 44
The Life of a Mealworm ... 47
The Invasion of the Lamprey ... 71
Eggs-amination ... 92
Comparing Feathers ... 104
Vertebrate Body Systems ... 121
Migration of Mammals ... 133

Activity: Doing
Mollusks in the Supermarket ... 30
Tunicates ... 66
Observing a Fish ... 72
A Frog Jumping Contest ... 79
Snakes ... 94
Endangered Mammals ... 127

Activity: Calculating
The Fastest Runner ... 130

Activity: Thinking
A Flock of Phrases ... 113

Activity: Writing
Symmetry ... 15
What Kind of Insect Is It? ... 52
Useful Mammals ... 129

Activity: Reading
A Silky Story ... 42
A Fish Story ... 74
Jumping Frogs ... 81
It All Depends on Your Point of View ... 91
Rara Aves ... 109
A Cat and Two Dogs ... 123

Problem Solving
Do You Want to Dance? ... 53
Alligator Anxieties ... 102

Connections
No Bones About It ... 23
Insects in Flight ... 57
Can Toads Cause Warts? ... 83
Flights of Fancy ... 113
Do You Hear What I Hear? ... 137

Careers
Veterinarian Assistant ... 131

CONCEPT MAPPING

hroughout your study of science, you will learn a variety of terms, facts, figures, and concepts. Each new topic you encounter will provide its own collection of words and ideas—which, at times, you may think seem endless. But each of the ideas within a particular topic is related in some way to the others. No concept in science is isolated. Thus it will help you to understand the topic if you see the whole picture; that is, the interconnectedness of all the individual terms and ideas. This is a much more effective and satisfying way of learning than memorizing separate facts.

Actually, this should be a rather familiar process for you. Although you may not think about it in this way, you analyze many of the elements in your daily life by looking for relationships or connections. For example, when you look at a collection of flowers, you may divide them into groups: roses, carnations, and daisies. You may then associate colors with these flowers: red, pink, and white. The general topic is flowers. The subtopic is types of flowers. And the colors are specific terms that describe flowers. A topic makes more sense and is more easily understood if you understand how it is broken down into individual ideas and how these ideas are related to one another and to the entire topic.

It is often helpful to organize information visually so that you can see how it all fits together. One technique for describing related ideas is called a **concept map**. In a concept map, an idea is represented by a word or phrase enclosed in a box. There are several ideas in any concept map. A connection between two ideas is made with a line. A word or two that describes the connection is written on or near the line. The general topic is located at the top of the map. That topic is then broken down into subtopics, or more specific ideas, by branching lines. The most specific topics are located at the bottom of the map.

To construct a concept map, first identify the important ideas or key terms in the chapter or section. Do not try to include too much information. Use your judgment as to what is really important. Write the general topic at the top of your map. Let's use an example to help illustrate this process. Suppose you decide that the key terms in a section you are reading are School, Living Things, Language Arts, Subtraction, Grammar, Mathematics, Experiments, Papers, Science, Addition, Novels. The general topic is School. Write and enclose this word in a box at the top of your map.

SCHOOL

Now choose the subtopics—Language Arts, Science, Mathematics. Figure out how they are related to the topic. Add these words to your map. Continue this procedure until you have included all the important ideas and terms. Then use lines to make the appropriate connections between ideas and terms. Don't forget to write a word or two on or near the connecting line to describe the nature of the connection.

Do not be concerned if you have to redraw your map (perhaps several times!) before you show all the important connections clearly. If, for example, you write papers for Science as well as for Language Arts, you may want to place these two subjects next to each other so that the lines do not overlap.

One more thing you should know about concept mapping: Concepts can be correctly mapped in many different ways. In fact, it is unlikely that any two people will draw identical concept maps for a complex topic. Thus there is no one correct concept map for any topic! Even

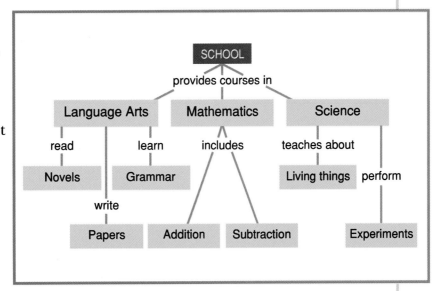

though your concept map may not match those of your classmates, it will be correct as long as it shows the most important concepts and the clear relationships among them. Your concept map will also be correct if it has meaning to you and if it helps you understand the material you are reading. A concept map should be so clear that if some of the terms are erased, the missing terms could easily be filled in by following the logic of the concept map.

Parade of Life: Animals

PARADE
OF LIFE
Animals

Everyone loves a friendly animal—the cricket that warms itself by a campfire, the frog that "sings" on a rock in a pond, the first robin to alight on your lawn in the spring, the squirrel that "plants" acorns from your oak tree, or the porpoise that frolics near a boat full of people. These creatures delight us because, unlike most animals, they seem not to be frightened by us.

However, there are animals that scurry away at the mere sound, smell, or sight of humans. Deer, for example, rapidly flee when they hear or smell an intruder. Cottontails use their hind legs to run when their large ears hear approaching danger. When frightened by our presence, an octopus emits an inky cloud to make good its escape!

Other animals—such as termites, flies, rats, mosquitoes, cockroaches, and hookworms—are regarded as pests. These animals cost us millions of dollars annually in damage to property and health. Still other animals are seen as terrifying menaces: the great white sharks, rattlesnakes, lions, tigers, and wolves.

Invertebrates, or animals without backbones, inhabit the Earth's land, sea, and air. The sponge (top) and the insect (right) are two examples of invertebrates.

8 ■ C

CHAPTERS

1 Sponges, Cnidarians, Worms, and Mollusks
2 Arthropods and Echinoderms
3 Fishes and Amphibians
4 Reptiles and Birds
5 Mammals

The wolf (left) belongs to a group of vertebrates, or animals with backbones, known as mammals. Like the wolf, the iguana (bottom) is a vertebrate. It belongs to a group of animals known as reptiles.

Whatever their behavior toward us, each of these animals has a specific role in nature. As you read this textbook, you will discover a great deal about the animals that inhabit the Earth's lands, water, and air. So read on and discover why the Earth is referred to as the living planet.

Discovery Activity

Animals, Animals

1. Take a look at a small area near your home. The area can be located in a park, an empty lot, a yard, or on a beach.

2. Make a list of all the animals you observe in the area. Make a sketch of each animal.

   ■ What are the characteristics of the animals you observed?

   ■ Which animals have similar characteristics? Which have different characteristics?

   ■ If you could develop your own system of classification, how would you group each animal on your list?

C ■ 9

covery Activity. Point out that students can organize their notes to help them set up their classification system. Have them think of questions they can use to compare and contrast the animals. For example, how many legs do they have? What are they covered with? How do they move? What do they eat? Suggest that students make a chart using their questions as headings for categories and then list the animals in the appropriate categories. Help students put their sketches and chart in a booklet. Ask volunteers to show their booklets to the class and explain their classification system. Explain that when students have completed the chapter, they may want to return to their booklets and add more information about the animals to their charts.

CHAPTER DESCRIPTIONS

1 Sponges, Cnidarians, Worms, and Mollusks Chapter 1 begins with a review of the five-kingdom classification system before focusing specifically on the animal kingdom. After a discussion of invertebrates and vertebrates, some of the earliest evolving groups of invertebrates are introduced. Form and function in sponges, cnidarians, worms, and mollusks are examined.

2 Arthropods and Echinoderms In Chapter 2 the general characteristics of arthropods, members of the largest animal phylum, are outlined. Then the structure, growth, development, behavior, and defense mechanisms of insects, the largest group in this phylum, are discussed. Finally, the chapter closes with information on the form and function of echinoderms.

3 Fishes and Amphibians Chapter 3 deals with the first groups of vertebrates to evolve—fishes and amphibians. In the first half of the chapter, the general characteristics and major types of fishes—including jawless, cartilaginous, and bony—are described. The second half of the chapter discusses the general characteristics and major types of amphibians.

4 Reptiles and Birds In the first half of Chapter 4, following a discussion of the basic characteristics of reptiles, the groups of reptiles—including lizards, snakes, turtles, alligators, and crocodiles—are described. In the second half of the chapter, the characteristics, behavior, and types of birds are examined.

5 Mammals Chapter 5 opens with an analysis of the main characteristics of mammals. Next, egg-laying and pouched mammals are described. Finally, the general characteristics of placental mammals and the more specific attributes of ten groups of these mammals are treated.

Chapter 1 SPONGES, CNIDARIANS, WORMS, AND MOLLUSKS

SECTION	HANDS-ON ACTIVITIES
1–1 The Five Kingdoms pages C12–C15 Multicultural Opportunity 1–1, p. C12 ESL Strategy 1–1, p. C12	**Student Edition** ACTIVITY BANK: To Classify or Not to Classify? p. C152 **Laboratory Manual** The Microscope: A Tool of the Life Scientist, p. C7 **Teacher Edition** Examining Plants and Animals, p. C10d
1–2 Introduction to the Animal Kingdom pages C15–C17 Multicultural Opportunity 1–2, p. C15 ESL Strategy, 1–2, p. C15	**Student Edition** ACTIVITY BANK: Friends or Foes? p. C153 **Laboratory Manual** Using the Microscope, p. C13
1–3 Sponges pages C18–C20 Multicultural Opportunity 1–3, p. C18 ESL Strategy 1–3, p. C18	**Student Edition** ACTIVITY (Discovering): Observing a Sponge, p. C19
1–4 Cnidarians pages C20–C25 Multicultural Opportunity 1–4, p. C20 ESL Strategy 1–4, p. C20	**Activity Book** ACTIVITY BANK: An Unusual Cnidarian, p. C137 **Laboratory Manual** Sponges and Hydras, p. C19
1–5 Worms pages C25–C30 Multicultural Opportunity 1–5, p. C25 ESL Strategy 1–5, p. C25	**Student Edition** ACTIVITY (Discovering): Worms at Work, p. C29 LABORATORY INVESTIGATION: Observing Earthworm Responses, p. C34 **Activity Book** ACTIVITY BANK: Flat as a Worm, p. C141 **Laboratory Manual** The Earthworm, p. C23 Flatworms and Roundworms, p. C31
1–6 Mollusks pages C30–C33 Multicultural Opportunity 1–6, p. C30 ESL Strategy 1–6, p. C30	**Student Edition** ACTIVITY (Doing): Mollusks in the Supermarket, p. C30 ACTIVITY BANK: Moving at a Snail's Pace, p. C154 **Laboratory Manual** Examining a Clam, p. C35
Chapter Review pages C34–C37	

OTHER ACTIVITIES	MEDIA AND TECHNOLOGY
Activity Book CHAPTER DISCOVERY: Discovering Symmetry, p. C9 **Review and Reinforcement Guide** Section 1–1, p. C5	**Interactive Videodisc/CD ROM** Amazonia **English/Spanish Audiotapes** Section 1–1
Student Edition ACTIVITY (Writing): Symmetry, p. C15 **Review and Reinforcement Guide** Section 1–2, p. C9	**Video** Animals Without Backbones (Supplemental) Adaptations of Animals (Supplemental) **English/Spanish Audiotapes** Section 1–2
Review and Reinforcement Guide Section 1–3, p. C13	**English/Spanish Audiotapes** Section 1–3
Review and Reinforcement Guide Section 1–4, p. C15	**English/Spanish Audiotapes** Section 1–4
Activity Book ACTIVITY: Comparing Digestive Systems of Worms, p. C15 **Review and Reinforcement Guide** Section 1–5, p. C19	**Video** Sponges, Coelenterates, and Worms (Supplemental) **English/Spanish Audiotapes** Section 1–5
Activity Book ACTIVITY: Mollusks in Our World, p. C13 ACTIVITY: Investigating the Behavior of an Octopus, p. C17 **Review and Reinforcement Guide** Section 1–6, p. C23	**Interactive Videodisc/ CD ROM** Virtual BioPark **English/Spanish Audiotapes** Section 1–6
Test Book Chapter Test, p. C9 Performance-Based Tests, p. C109	**Test Book** Computer Test Bank Test, p. C15

*All materials in the Chapter Planning Guide Grid are available as part of the Prentice Hall Science Learning System.

CHAPTER OVERVIEW

There are billions of living things on the Earth. These billions are divided into millions of separate kinds with individual characteristics and structures. It would be impossible to study each living thing without some workable system of classification. The system most commonly used today places organisms into five kingdoms: monerans, protists, fungi, plants, and animals. The kingdoms are subdivided into phyla; phyla are subdivided into smaller groups called classes; classes into orders; orders into families; families into genera; and genera into species. Each plant and animal is given two scientific names: a genus name and a species name.

The animal kingdom is divided into two main sections: vertebrates and invertebrates. Invertebrate phyla include sponges, cnidarians, flatworms, roundworms, annelids, and mollusks.

The simplest invertebrates are the sponges, which are little more than colonies of cells living together. Their bodies remain stationary as their cells filter food and oxygen from the water entering their pores. Somewhat more advanced are the cnidarians. Unlike sponges, their body cells are organized into true tissues.

Worms are grouped into three phyla on the basis of their body plan. Flatworms have ribbonlike bodies and an incomplete digestive tract, whereas roundworms are threadlike and have a digestive tract with two openings. Many flatworms and roundworms are parasites. Annelids have rounded bodies with segments. With segmentation, a greater specialization of body parts is possible.

The majority of mollusks are aquatic creatures. Mollusks include such diverse animals as snails, clams, and squids. But despite their diversity, they all have soft bodies covered by a mantle that often secretes a limy shell.

1–1 THE FIVE KINGDOMS
THEMATIC FOCUS

The purpose of this section is to introduce students to the classification system used for living organisms. The five kingdoms are described, and examples are given of organisms in each kingdom.

The themes that can be focused on in this section are evolution and unity and diversity.

*Evolution: Organisms are classified by similarities in structure, function, and evolutionary heritage.

*Unity and diversity: Although each living organism is unique, it may be grouped with others of its kind for purposes of scientific study.

PERFORMANCE OBJECTIVES 1–1

1. List the five kingdoms used to classify all life on Earth.
2. Describe some characteristics of each kingdom.

SCIENCE TERMS 1–1

kingdom p. C12
autotroph p. C12
heterotroph p. C13

1–2 INTRODUCTION TO THE ANIMAL KINGDOM
THEMATIC FOCUS

The purpose of this section is to show students that the animal kingdom can be divided into two main groups: vertebrates and invertebrates. Vertebrates have a backbone; invertebrates do not. Many people automatically think of vertebrates when hearing the word *animal,* but the invertebrates are by far the more numerous, making up more than 90 percent of all animal species.

The themes that can be focused on in this section are systems and interactions and energy.

Systems and interactions: Invertebrates respond to and interact with their environment in ways that help them gather food, protect themselves, and reproduce.

Energy: All animals are heterotrophs. They obtain energy by eating autotrophs or other heterotrophs.

PERFORMANCE OBJECTIVES 1–2

1. Define vertebrate and give some examples.
2. Define invertebrate and give some examples.

SCIENCE TERMS 1–2

vertebrate p. C15
invertebrate p. C15

1–3 SPONGES
THEMATIC FOCUS

The purpose of this section is to introduce students to sponges, among the simplest of the invertebrates. Unlike most animals, these aquatic organisms usually remain attached to one spot their entire lives. Their bodies contain numerous pores through which food and oxygen enter the organisms. The name of the phylum, Porifera, comes from the Latin word *porifera,* meaning pore-bearer.

The themes that can be focused on in this section are systems and interactions and scale and structure.

Systems and interactions: Even though very simple in organization, sponges have all the necessary functions needed to sustain life.

*Scale and structure: The organs and other body structures that carry out an animal's life functions vary from phylum to phylum.

PERFORMANCE OBJECTIVES 1–3

1. Describe the physical appearance of a sponge.
2. Explain the methods by which sponges obtain food and oxygen.
3. Explain the difference between sexual and asexual reproduction.

SCIENCE TERMS 1–3

spicules p. C19
sexual reproduction p. C19
asexual reproduction p. C20

1-4 CNIDARIANS
THEMATIC FOCUS

The purpose of this section is to survey the cnidarians, a group of invertebrates characterized by a central cavity with only one opening and with stinging tentacles surrounding that opening. Most cnidarians live in salt water, and many are strikingly beautiful. Familiar cnidarians include corals, jellyfish, and sea anemones.

The themes that can be focused on in this section are unity and diversity and patterns of change.

*Unity and diversity: Different groups of invertebrates carry out the basic life functions in a variety of ways.

*Patterns of change: The levels of organization become higher as animals become more complex.

PERFORMANCE OBJECTIVES 1-4

1. Describe the characteristics of cnidarians.
2. Give examples of organisms classified as cnidarians.

SCIENCE TERMS 1-4

nematocyst p. C21

1-5 WORMS
THEMATIC FOCUS

The purpose of this section is to introduce students to three main types of worms. The least complex are the flatworms, of which there are both parasitic and free-living forms. Among the free-living flatworms are the planarians, which are found living in ponds, lakes, streams, and oceans.

Roundworms have not only a rounded body, but also a complete digestive system with two openings—a mouth at the anterior end and an anus at the posterior end. Roundworms include many parasitic forms, such as *Trichinella,* flukes, and hookworms.

Segmented worms have bodies made up of many sections. These worms are found in oceans and freshwater streams and lakes, but the most familiar segmented worm, the earthworm, is found in the soil.

The themes that can be focused on in this section are unity and diversity and stability.

*Unity and diversity: Although worms live in many different habitats and exhibit marked differences in size and behavior, all worms share common characteristics that allow them to be grouped together in phyla.

Stability: Although the levels of organization become higher as animals become more complex, all animals maintain a stable internal and external environment.

PERFORMANCE OBJECTIVES 1-5

1. Name the three groups of worms and give examples of each.
2. Compare the shapes and body plans of the three groups of worms.

SCIENCE TERMS 1-5

regeneration p. C26
parasite p. C26
host p. C27

1-6 MOLLUSKS
THEMATIC FOCUS

The purpose of this section is to introduce students to the mollusks, a group of invertebrates having a soft, unsegmented body. Three major groups of mollusks are discussed in this section: gastropods such as snails and slugs; bivalves such as clams, mussels, and scallops; and the tentacled mollusks such as the octopus, squid, and nautilus.

The themes that can be focused on in this section are scale and structure and evolution.

*Scale and structure: The mollusks include a wide variety of forms that are classified according to the presence of a shell, the type of shell, and the type of foot.

*Evolution: All invertebrates share an evolutionary heritage. The remarkably well-preserved fossils of early mollusks have provided significant insights into the evolution of early organisms.

PERFORMANCE OBJECTIVES 1-6

1. Describe the characteristics of mollusks.
2. identify the three major groups of mollusks and give examples of each.

Discovery *Learning*

TEACHER DEMONSTRATIONS MODELING

Examining Plants and Animals

Set up a display around the classroom of some common specimens of plants and animals. Some possibilities:

1. algae floating in water
2. clump of moss
3. *Elodea*
4. potted geranium
5. potted ivy
6. potted fern
7. branch from a tree
8. earthworm
9. snail
10. cricket
11. crayfish
12. fish in an aquarium
13. mouse

If you cannot obtain living specimens, then display a variety of pictures of plants and animals or a combination of specimens and pictures. Allow the class ample time to move about the room to examine the specimens or pictures. After students return to their seats, conduct a discussion on some of the similarities and differences between plants and animals.

• **If you were to place the organisms you just looked at into two different groups, how would you do it?** (Answers may vary. Lead students to suggest that some are plants and the others are animals.)
• **How do animals obtain their food?** (Answers will vary. Develop the idea that animals are ingestors of food. Most of them take in food and digest it internally.)

CHAPTER 1

Sponges, Cnidarians, Worms, and Mollusks

INTEGRATING SCIENCE

This life science chapter provides you with numerous opportunities to integrate other areas of science, as well as other disciplines, into your curriculum. Blue numbered annotations on the student page and integration notes on the teacher wraparound pages alert you to areas of possible integration.

In this chapter you can integrate life science and classification (pp. 12, 16), life science and ecology (p. 12), life science and botany (p. 14), life science and cell biology (p. 15), life science and evolution (p. 18), language arts (pp. 18, 21, 26, 30), earth science and reef formation (p. 22), life science and symbioses (pp. 23, 26), life science and the skeletal system (p. 23), food science (p. 27), and mathematics (p. 33).

SCIENCE, TECHNOLOGY, AND SOCIETY/COOPERATIVE LEARNING

Many of the annelids known as leeches are external parasites that drink the blood and body fluids of their hosts. Specialized secretions in leech saliva prevent the hosts' blood from clotting and anesthetize the wound so that the unlucky hosts will not know that they have been attacked.

Leeches were used by doctors as early as 200 BC as a medical treatment to restore health. It was believed that by "bleeding" the patient with leeches, the "bad blood" causing the illness would be removed and the patient would be healed. As doctors learned that microorganisms, not "bad blood," cause disease, the use of leeches was all but discontinued.

Today, leeches and the chemicals in leech saliva are again being used by doctors to treat patients. In combination with other "clot-busting" drugs, hirudin (the first antithrombin extracted from leech saliva) is expected to enhance the ability of clot-busters to destroy clots in

INTRODUCING CHAPTER 1

DISCOVERY LEARNING

▶ *Activity Book*

Begin your introduction to this chapter by using the Chapter 1 Discovery Activity from the *Activity Book*. Using this activity, students will investigate the various types of symmetry found in living organisms.

USING THE TEXTBOOK

Ask the class to look at the photograph and read the chapter-opening paragraphs.
• **What animal is shown in the photograph?** (An octopus.)
• **How does this animal move through the water?** (It pulls itself over rocks with its arms or swims by squirting out jets of water.)
• **How many arms or tentacles does an octopus have?** (Eight; the prefix *octo-* indicates eight.)

Sponges, Cnidarians, Worms, and Mollusks

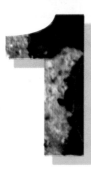

Guide for Reading

After you read the following sections, you will be able to

1–1 The Five Kingdoms
- Describe the five kingdoms of living things.

1–2 Introduction to the Animal Kingdom
- Compare vertebrates and invertebrates.

1–3 Sponges
- Describe the characteristics of sponges.

1–4 Cnidarians
- Describe the characteristics of cnidarians.

1–5 Worms
- Compare flatworms, roundworms, and segmented worms.

1–6 Mollusks
- Identify three groups of mollusks.

No other beast of the sea has been as feared as the octopus. Its very name—devilfish—calls to mind danger and alarm. Yet the truth is that the octopus would be more afraid of you than you should be of it. The octopus, you see, is terrified of anything larger than itself.

Of the hundred or so varieties of octopuses, most grow no larger than one meter across. Some are so tiny that they could sit on your fingernail! In the Mediterranean Sea, where octopuses are common, very few grow tentacles that reach two meters in length. Only in the depths of the Pacific Ocean are there believed to be really tremendous octopuses.

No matter what their actual size, the tales told about octopuses were common among people with vivid imaginations. In such stories, octopuses attacked ships, slung their tentacled arms around the masts, and nearly capsized the vessels. Actually, few scientists believe that octopuses intentionally attack humans.

Octopuses are one of many fascinating animals you will read about in this textbook. In this chapter you will begin to explore this marvelous world by examining sponges, cnidarians, worms, and mollusks.

Journal *Activity*

You and Your World In your journal, write down five common expressions that have to do with animals. Here are two examples: "At a snail's pace" and "It's a fluke." Next to each expression, explain its meaning as you understand it. When you have finished reading this textbook, look again at your list. See if you can explain the origin of each expression.

An octopus resting on its tentacles

---

the coronary arteries. And to prevent coronary arteries opened by angioplasty from closing again, doctors plan to use hirudin to keep these arteries open, thus eliminating the need for additional surgical procedures.

Cooperative learning: Using preassigned lab groups or randomly selected teams, have groups complete one of the following assignments:
- Write an "update letter" to the unknown Greek physician who first recorded the use of leeches in medical treatment. Students should invent an imaginary name and address for the physician. Their letters should explain the history of leeching since his death and should describe how leeches are being used in medicine today.
- Prepare a 30-second item for the evening news describing the role of leeches in medicine. The news item should describe leeches and how they obtain their food and should contain information on early uses of leeches by doctors and how leeches are being used today. Along with a script, groups should also include visuals that they would use in the newscast.

See Cooperative Learning in the *Teacher's Desk Reference.*

JOURNAL ACTIVITY

You may want to use the Journal Activity as the basis of class discussion. After students have listed their expressions, lead a class activity in which the animals are divided into vertebrates (those with backbones) and invertebrates (those without backbones). Students should be instructed to keep their Journal Activity in their portfolio.

Explain that a squid is an animal similar to an octopus. Squids, however, have ten arms rather than eight.
- **Do you think these animals are dangerous?** (Accept all answers, but ask students to substantiate their opinions.)

Point out that even though large octopuses and squids can be dangerous if provoked, there are no confirmed records of their being naturally aggressive toward humans. Most of them are small, shy animals that slip away quietly in the presence of people. Despite legends and the imagination of movie directors, there has never been a single case of a diver being trapped by a giant octopus.
- **Would you guess that octopuses are very intelligent?** (Accept all answers.)

Explain that octopuses are considered the most intelligent of the invertebrates. Invertebrates are animals without backbones. In laboratory studies, octopuses learn quickly and can remember for periods of up to several weeks.

- **What is the main difference between octopuses and animals such as dogs, turtles, rabbits, and humans?** (Accept all answers, but lead students to understand that the latter animals have backbones.)
- **Why do you think story and movie writers use animals such as octopuses and squids as the "villains" in their plots?** (Accept all answers. Students may guess that more unfamiliar animals would make more believable villains.)

1-1 The Five Kingdoms

MULTICULTURAL OPPORTUNITY 1-1

Point out that artists and sculptors throughout the world have frequently depicted animals. Although animals are often portrayed because of their form or attractive colors, they are also used as symbols. In African art, for instance, animals often represent abstract ideas. Strength may be symbolized by buffaloes, crocodiles, elephants, and lions. Swiftness of movement is symbolized by snakes, the tortoise may represent old age, and life may be symbolized by the lizard. Ask students to investigate the symbols that animals represent in the arts and folklore of various cultures.

ESL STRATEGY 1-1

Have students complete the names of the five kingdoms of living things and then rearrange them in order from the simplest to the most complex.

F _____
M _____
P _____
A _____
P _____

(**Answers:** Monera, Protista, Fungi, Plantae, Animalia.)

For a better understanding of the terms *autotroph* and *heterotroph*, have students look up in a dictionary the meaning of the prefixes *auto-* and *hetero-*. Ask them to think of other words that begin with these prefixes. Explain that the suffix *-troph* comes from a Greek word meaning "nourishment."

Guide for Reading

Focus on this question as you read.

▶ *What are some general characteristics of each of the five kingdoms?*

Ａctivity Bank

To Classify or Not to Classify?, p.152

Figure 1-1 *The many different types of environments in which organisms live include tropical rain forests and ocean depths.*

12 ■ C

1-1 The Five Kingdoms

Just by looking around you, you can see that you share the Earth with many different types of organisms, or living things. Insects, cows, horses, trees, flowers, grasses, bacteria, mushrooms, and fishes are but a few examples. But what you may not know is that scientists have already identified more than 2.5 million species, or groups of organisms that share similar characteristics and that can breed with one another. What is even more amazing is that the job is far from over! In fact, some scientists estimate that there may be millions more organisms living in areas such as the tropical rain forests and the lower depths of the Earth's oceans that have not as yet been identified.

To help study all known species, which is an impossible task at best, scientists have divided all life on Earth into several **kingdoms,** or large, general groups. **Today, the most generally accepted system of classification contains five kingdoms: monerans** (MOHN-er-ans), **protists** (PROHT-ihsts), **fungi** (FUHN-jigh; singular: fungus), **plants, and animals.** ❶

As is often the case in science, not all scientists agree on this classification system. And their disagreement is important for you to keep in mind. Someday, research may show that different classification systems make more sense and better describe how living things have evolved (changed). But for now, this five-kingdom classification system is a useful way to study living things.

MONERANS All the Earth's bacteria are members of the kingdom Monera. Monerans are unicellular organisms, or organisms that are made of one cell. The cell of a moneran does not contain a nucleus, which is the control center of a cell. A moneran's cell, in fact, does not have many of the structures found in other cells. This fact has led scientists to believe that monerans are distant relatives of the other four kingdoms.

Like other organisms, monerans can be placed into two divisions, based on how they get their energy. Organisms that obtain their energy by making their own food are called **autotrophs** (AW-toh-trohfs). Organisms that cannot make their own food ❷

TEACHING STRATEGY 1-1

FOCUS/MOTIVATION

Have students examine the photographs on pages C12 and C13. Ask for volunteers to read the captions.
• **How could the organisms shown in the photographs be divided into groups?** (According to similar traits.)
• **What information about the evolutionary relationships among organisms** is revealed by studying the traits of organisms? (Similar structures in organisms indicate similar evolutionary ancestors.)

🎞 Media and Technology

Have students view the Interactive Video/CD ROM entitled Amazonia and have them classify any five of the organisms that they observe into their appropriate kingdom. In subsequent sections and chapters, students can continue to explore and discover the unity and diversity among organisms.

CONTENT DEVELOPMENT

Develop the idea that dividing organisms into groups aids the investigation of Earth's life forms. It is important, however, for scientists worldwide to use similar methods of classification.
• **Why is it necessary for scientists worldwide to use a similar grouping system?** (A universally accepted clas-

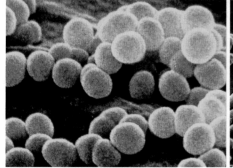
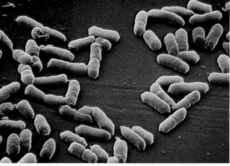

are called **heterotrophs** (HEHT-er-oh-trohfs). Heterotrophs may eat autotrophs in order to obtain food. Or they may eat other heterotrophs. Whatever the case, all heterotrophs rely on autotrophs for food.

Scientists have evidence that monerans first appeared on Earth about 3.5 billion years ago. (Remember, the Earth is estimated to be more than 4 billion years old.) This makes monerans the earliest life forms on Earth.

PROTISTS The kingdom Protista contains most of the unicellular organisms that have a nucleus. In addition to a nucleus, protists also have special cell structures that perform specific functions.

Some protists move like animals but at the same time have several obvious plantlike characteristics. Specifically, they are green in color and can use light energy to make their own food from simple substances.

Protists were the first living things to contain a nucleus. Ancient types of protists that lived millions of years ago are probably the ancestors of modern protists, fungi, plants, and animals.

FUNGI Fungi make up the kingdom Fungi. Most fungi are multicellular organisms, or organisms that are made of many cells. Although you may not realize it, you are probably quite familiar with fungi. Mushrooms are fungi. So are the molds that sometimes grow on leftover food that has stayed too long in the refrigerator. And the mildews that form small black spots in damp bathrooms and basements are also fungi.

Figure 1–2 These spherical-shaped bacteria and rod-shaped bacteria are placed in the kingdom Monera. What are the characteristics of monerans? 1

Figure 1–3 The kingdom Protista contains most of the unicellular organisms that have a nucleus. Examples of protists include Trichomonas.

C ■ 13

• **Does only the fact that an organism cannot make its own food mean it is an animal?** (No.)
• **Can you think of some organisms that cannot make their own food but are not animals?** (Fungi, some protists, and bacteria.)
• **What are cell walls?** (Outermost boundary of most cells, with the exception of animal cells.)
• **What organisms have cell walls?** (Plants, fungi, bacteria, some protists.)
• **Because they have cell walls, these organisms are not animals. What other characteristics have we discussed that indicate that these organisms are not animals?** (Plants are autotrophs.)

● ● ● ● **Integration** ● ● ● ●

Use the discussion of autotrophs and heterotrophs to integrate concepts of ecology into your lesson.

GUIDED PRACTICE

▶ *Laboratory Manual*

Skills Development

Skills: Interpreting diagrams, making observations, making inferences

At this point you may want to have students complete the Chapter 1 Laboratory Investigation in the *Laboratory Manual* called The Microscope: A Tool of the Life Scientist. In the investigation students will learn to identify the parts of a microscope and examine a prepared slide.

sification system eliminates confusion and permits scientists to understand one another's data.)

Develop the idea that a particular organism may have different English names but only one scientific name. The scientific name for a panther and a leopard is *Panthera pardus*.
• **What does this indicate about a panther and a leopard?** (They are the same animal.)

● ● ● ● **Integration** ● ● ● ●

Use the discussion of the five kingdoms to integrate concepts of classification into your lesson.

CONTENT DEVELOPMENT

Explain the difference between autotrophs and heterotrophs. Autotrophs can make their own food and heterotrophs cannot.
• **What living organisms are autotrophs?** (Plants, some bacteria, and protists.)

ACTIVITY
WRITING
SYMMETRY

In animals with radial symmetry, body parts repeat around an imaginary line drawn through the center of their bodies. These animals never have any kind of a real "head." In animals with bilateral symmetry, body parts repeat on either side of the imaginary center line. Asymmetrical animals do not have any type of symmetry.

Most sponges and cnidarians have radial symmetry. Most worms and mollusks have bilateral symmetry.

1–1 (continued)

CONTENT DEVELOPMENT

Point out to students that animals share certain characteristics with plants. Perhaps the most significant of these in both vascular plants and in higher animals is the organization of cells into specialized tissues.

● ● ● ● Integration ● ● ● ●

Use the discussion of the plant kingdom to integrate concepts of botany into your lesson.

INDEPENDENT PRACTICE
Section Review 1–1

1. Monerans, protists, fungi, plants, and animals.

Figure 1–4 *The red mushroom is a fungus. Can you name other examples of fungi?* 1

Until a short time ago, fungi were classified as plants. However, fungi differ from plants in several basic ways. The most obvious difference between plants and fungi is that plants are able to use light energy to make their own food from simple substances. Fungi are not. Fungi, like animals, must obtain their energy from another source.

PLANTS As you might expect, plants make up the kingdom Plantae. Most members of the plant kingdom are multicellular organisms. The members of this kingdom that are probably most familiar to you are flowering plants, trees, mosses, ferns, and some algae.

ANIMALS Animals are the multicellular organisms that make up the kingdom Animalia. Animals have specialized tissues, and most have organs and organ systems. Because animals cannot make their own food, they are heterotrophs. And, unlike plant cells, animal cells do not have cell walls.

Figure 1–5 *The two kingdoms of organisms that you are probably most familiar with are the plants and the animals. Most plants, such as the trees and mosses on this forest floor, are multicellular (left). However, there are some plants that are unicellular. All animals, such as the hippopotamus (center) and the helmeted lizard (right), are multicellular organisms.*

14 ■ C

---

2. Monerans are unicellular but do not contain nuclei. Protists are unicellular with nuclei. Most fungi are multicellular and cannot make their own food. Plants can make their own food, are multicellular, and have cells with cell walls. Animals are multicellular, cannot make their own food, and their cells lack cell walls.

3. Autotrophs can make their own food. Heterotrophs must get their food by eating autotrophs or other heterotrophs.

4. Probably the animal kingdom—it could not belong to the monerans or the protists because it is multicellular; it could not be a plant or a fungus because it does not have cell walls.

REINFORCEMENT/RETEACHING

Review students' responses to the Section Review questions. Reteach any material that is still unclear, based on students' responses.

1–1 Section Review

1. What are the five kingdoms of living things?
2. List two important characteristics of each kingdom.
3. How does an autotroph obtain its food? A heterotroph?

Critical Thinking—*Applying Concepts*

4. Suppose a new species is found to be a multicellular heterotroph whose cells lack cell walls. To which kingdom of organisms would this species belong? Explain.

1–2 Introduction to the Animal Kingdom

Think of a fierce lion, a friendly porpoise, a cuddly puppy, a crawling earthworm, an annoying mosquito, and a slimy jellyfish. Now ask yourself what all these organisms have in common. You are correct if you say they are all animals and belong to the kingdom Animalia. As you have just learned, animals can be defined as multicellular heterotrophs whose cells lack cell walls.

Animal cells vary greatly in size. Although most cells are too small to be seen without a microscope, some cells are large enough to be seen with the unaided eye. The largest animal cell is the yolk of an ostrich egg. This yolk, or single cell, is about the size of a baseball. Animal cells also have different shapes. Some are shaped like long rectangles, some like spheres, some like disks. Others are rod shaped or spiral shaped.

In most animals, cells are organized into tissues, tissues into organs, and organs into organ systems. Every kind of cell depends on every other kind of cell for its survival and for the survival of the entire animal.

The animal kingdom can be grouped into two major divisions: **vertebrates** and **invertebrates.**

Guide for Reading

Focus on this question as you read.

▶ What is the difference between a vertebrate and an invertebrate?

ACTIVITY
WRITING

Symmetry

The body shapes of invertebrates show either radial symmetry, bilateral symmetry, or asymmetry. Use a science dictionary or science encyclopedia to define each term.

Make a list of the different phyla of invertebrates discussed in this chapter. Indicate what type of symmetry is shown by each phylum.

C ■ 15

1–2 Introduction to the Animal Kingdom

MULTICULTURAL OPPORTUNITY 1–2

Suggest that students find out about the distinguished African-American artist Romare Bearden (1911–1988), whose paintings and collages often depict animals, birds, reptiles, and fishes. Animals are used to represent the artist's viewpoint and reflect his cultural heritage. Interested students may want to examine some of the artist's paintings, for example, *The Indigo Snake,* and try to identify the organisms used in each work.

ESL STRATEGY 1–2

Have students complete this analogy.

car models : car factory
phyla : _____

Tell students that the word that goes in the blank should relate to the word next to it in the same way as the words in the example relate to each other. Ask students to give reasons for their choice. (**Answer:** kingdom.)

CLOSURE

▶ *Review and Reinforcement Guide*
Have students complete Section 1–1 in the *Review and Reinforcement Guide.*

TEACHING STRATEGY 1–2

FOCUS/MOTIVATION

Obtain three or more pictures of invertebrates and a similar number of pictures of vertebrates. Have students compare the pictures.

• **Can you think of a way in which we can divide these animals into two groups?** (Accept all logical answers, but lead students to suggest that some of the pictured animals have a backbone and some do not.)
• **What do we call animals that have a backbone?** (Animals with a backbone are called vertebrates.)
• **What do we call animals without backbones?** (Animals without backbones are called invertebrates.)

CONTENT DEVELOPMENT

Ask students to try to write down the name of one invertebrate and one vertebrate. Place two columns headed Invertebrates and Vertebrates on the chalkboard or overhead projector and then call on students one at a time to name their examples. List examples in the appropriate column as they are given.

● ● ● ●   **Integration**   ● ● ● ●

Use the discussion of animal cells to integrate concepts of cell biology into your lesson.

To date, scientists have identified more than 2.5 million species of organisms on Earth. Such great diversity among living things called for a system of naming and ordering the organisms in a logical manner. In the eighteenth century, such a system was devised by a Swedish botanist named Carolus Linnaeus.

Linnaeus's system of naming organisms is called binomial nomenclature. In this system, each organism is assigned a two-part scientific name that identifies its genus and species.

Linnaeus grouped organisms according to similar body structures. The groups to which organisms are assigned are called taxa. Thus, the science of naming and grouping living things is called taxonomy. The taxa that make up Linnaeus's classification system are kingdom, phylum, class, order, family, genus, and species.

Linnaeus's two-kingdom classification system has evolved to a five-kingdom classification system widely used today. The five-kingdom system more accurately groups organisms according to evolutionary trends.

Figure 1–6 *All members of the kingdom Animalia can be grouped into two major divisions: vertebrates and invertebrates. Warthogs (top left) and dusky dolphins (top right) are vertebrates; sea stars (bottom left) and tarantulas (bottom right) are invertebrates. How can you distinguish between a vertebrate and an invertebrate?* ❶

Activity Bank

Friends or Foes?, p.153

Humans, lions, porpoises, and puppies are all examples of vertebrates. **A vertebrate is an animal that has a backbone, or vertebral** (VER-tuh-bruhl) **column.** Earthworms, mosquitoes, and jellyfishes are all invertebrates. **An invertebrate is an animal that has no backbone.** Invertebrates make up about 95 percent of all animal species.

❶ All animals in the kingdom Animalia are divided into different groups called phyla (FIGH-luh; singular: phylum) according to their body structure. A phylum is the second largest group of organisms after kingdom. As you explore the invertebrate phyla, keep in mind that they share an evolutionary heritage. In

---

1–2 (continued)

CONTENT DEVELOPMENT

Emphasize that even though vertebrates are generally larger and more familiar than invertebrates to most people, they are in the minority. No one really knows the number of animal species that exists, but some biologists feel there may be as many as 2 million. Among these, fewer than 50,000 species are verte-

brates. Point out that the insects alone, which are just one kind of invertebrate, outnumber all the vertebrates.

● ● ● ● **Integration** ● ● ● ●

Use the discussion of the phyla in the animal kingdom to integrate concepts of classification into your lesson.

CONTENT DEVELOPMENT

Explain that the phyla in the animal kingdom all share an evolutionary heri-

tage. Have students study the diagram in Figure 1–7.

● **Which of the animal pictures show vertebrates?** (Students should be able to identify the fish, frog, turtle, duck, and squirrel.)

Explain that all the other pictures are of invertebrates. Point out that the phylum to which vertebrates belong also contains some invertebrates.

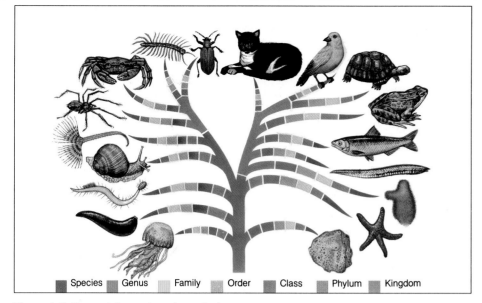

| Species | Genus | Family | Order | Class | Phylum | Kingdom |

Figure 1–7 *The evolutionary tree shows the best understanding of the evolutionary relationships between different groups of organisms. Which invertebrates seem to be most closely related to vertebrates?* ❷

other words, the invertebrates (as well as all animals) share a common ancestor. This fact will become evident to you as you move from one animal phylum to another.

1–2 Section Review

1. What is an animal?
2. How are vertebrates and invertebrates alike? How are they different?
3. What is a phylum? Would you expect to find more species in a phylum or a kingdom? Explain.

Critical Thinking—*Drawing Conclusions*
4. Why do multicellular organisms need specialized tissues?

C ■ 17

lum because phyla are subsumed in kingdoms. Because the classification system is based on characteristics rather than number of species, however, it is quite possible for a very large phylum such as the arthropods to have more species than one of the kingdoms.
4. Accept all logical responses. Students may suggest that the individual cells in multicellular organisms would be unable to carry out all the many different functions needed for the organism's survival.

REINFORCEMENT/RETEACHING
Review students' responses to the Section Review questions. Reteach any material that is still unclear, based on students' responses.

CLOSURE
▶ *Review and Reinforcement Guide*
Have students complete Section 1–2 in the *Review and Reinforcement Guide*.

GUIDED PRACTICE
▶ *Laboratory Manual*
Skills Development
Skills: Making observations, making comparisons
At this point you may want to have students complete the Chapter 1 Laboratory Investigation in the *Laboratory Manual* called Using the Microscope. In the investigation students will learn how to prepare a wet-mount slide.

INDEPENDENT PRACTICE
Section Review 1–2
1. A multicellular heterotroph whose cells lack cell walls.
2. Vertebrates and invertebrates lack cell walls and are heterotrophs. Vertebrates have backbones; invertebrates do not.
3. A phylum is a classification group for living organisms. Several phyla make up a kingdom.
Most students would expect to find more species in a kingdom than in a phy-

1-3 Sponges

MULTICULTURAL OPPORTUNITY 1-3

Bring to class a natural sponge and a synthetic sponge. Mention that in areas such as Greece, for instance, the once-important sponge-diving industry has declined because of competition from synthetic products. Ask students to think about why synthetic sponges were developed. What are some advantages—either functionally or in terms of the environment—of the synthetic sponge?

ESL STRATEGY 1-3

When explaining that pores are one of the main characteristics of sponges, remind students that there are many homonyms in English (words that sound the same but may be spelled differently and have different meanings). *Pore* and *pour* are homonyms. Have students use *pore* and *pour* in sentences.

To help students understand the term *asexual,* explain that many words beginning with the prefix *a-* are of Greek origin; this prefix means "not," "without," or "lacking." Have students think of other words with this prefix and explain what they mean.

TEACHING STRATEGY 1-3

FOCUS/MOTIVATION

If available, show the class some preserved specimens and/or pictures of sponges. Also have students look at Figure 1–8.

• **Why do you think people sometimes mistake sponges for plants?** (Accept all logical answers. Students will likely indicate that sponges are attached to one spot like most plants.)

• **Why are sponges classified as animals?** (They cannot carry on photosynthesis. They catch food and digest it inside their bodies.)

• **Why can sponges be used in bathing or in washing dishes?** (Some sponges have a soft, flexible, absorbent skeleton.)

Figure 1–8 *Sponges are invertebrates and belong to the phylum Porifera. As you can see from the photographs, sponges—such as* Callyspongia *sponges (left), tennis ball sponges (top), red sponges (bottom left), and Caribbean reef sponges (bottom right)—vary in shape, color, and size. What does the word porifera mean?* ❶

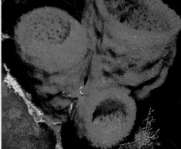
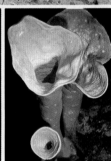

18 ■ C

1-3 Sponges

The simplest group of invertebrates are the sponges. Sponges are the most ancient of all animals ❶ alive today. The first sponges are thought to have appeared on Earth about 580 million years ago.

Although most sponges live in the sea, some can be found in freshwater lakes and streams. Sponges grow attached to one spot and usually stay there for their entire life unless a strong wave or current washes them somewhere else. Because they show little or no movement, sponges were once thought to be plants.

The body of a sponge is covered with many pores, or tiny openings. Sponges belong to the phylum Porifera (poh-RIHF-er-uh). The word porifera ❷ means pore-bearers. Moving ocean water carries food and oxygen through the pores into the sponge. The sponge's cells remove food and oxygen from the water. At the same time, the cells release waste products (carbon dioxide and undigested food) into the water. Then, the water leaves through a larger opening.

CONTENT DEVELOPMENT

Tell students that most sponges live in warm ocean waters near the coast but that there are also a few freshwater species. Explain that sponges are among the most ancient of all animals.

● ● ● ● **Integration** ● ● ● ●

Use the discussion of the first sponges to integrate concepts of evolution into your lesson.

GUIDED PRACTICE

Skills Development
Skill: Interpreting diagrams

Have students examine Figure 1–9. Point out the pores shown in the diagram.

• **What is the function of the pores?** (Food and oxygen are carried by moving water through the pores into the sponge.)

Point out that a sponge body has two

Sponge cells are unusual in that they function on their own, without any coordination with one another. In fact, some people think of sponges as a colony of cells living together. Despite their independent functioning, however, sponge cells have a mysterious attraction to one another. This attraction can be easily demonstrated by passing a sponge through a fine filter so that it breaks into clumps of cells. Within hours, these cells reform into several new sponges. No other animal species shares this amazing ability of sponge cells to reorganize themselves.

Many sponges produce **spicules** (SPIHK-yoolz). Spicules are thin, spiny structures that form the skeleton of many sponges. Spicules are made of either a chalky or a glasslike substance. They interlock to form delicate skeletons, such as the one shown in Figure 1–11 on page 20. Other sponges have skeletons that consist of a softer, fiberlike material. The cleaned and dried skeletons of these sponges are the natural bath sponges that you may have in your home or see in department stores. Still other sponges have skeletons that are made of both spicules and the fiberlike material.

Sponges reproduce sexually and asexually. **Sexual reproduction** is the process by which a new organism forms from the joining of a female cell and a male cell. Reproducing sexually, one sponge produces eggs (female cells); another produces sperm

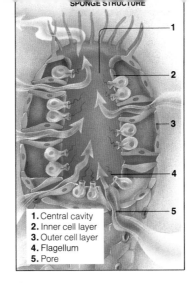

SPONGE STRUCTURE

1. Central cavity
2. Inner cell layer
3. Outer cell layer
4. Flagellum
5. Pore

Figure 1–9 *Notice that the body of a sponge consists of a layer of cells that form a wall around a central cavity. The cells in this layer function independently of one another.*

ACTIVITY
DISCOVERING

Observing a Sponge

1. Obtain a natural sponge from your teacher.
2. Use a hand lens to examine the surface and the pores. Draw what you see.
3. Remove a small piece of the sponge and place it on a glass microscope slide. Look for the spicules. Draw what you observe.

■ Repeat steps 2 and 3 with a synthetic kitchen sponge. How do a natural sponge and a synthetic sponge compare?

Figure 1–10 *In some sponges, such as the sponges in the photograph, eggs are squirted into the surrounding water, where they may be fertilized. In others, eggs are fertilized inside the body of the parent sponge.*

ACTIVITY
DISCOVERING
OBSERVING A SPONGE

Discovery Learning

Skills: Making comparisons

Materials: natural and synthetic sponges, hand lens, microscope

Most natural sponges have a mineral skeleton made up of tiny, needlelike spicules. The spicules may be either calcium carbonate (limestone) or silica, a glasslike material. In bath sponges, the skeleton consists of fibers of tan-colored protein called spongin. The skeleton of spongin fibers is what remains after a bath sponge dies and its cells are removed. Many sponges have a skeleton of both mineral spicules and spongin fibers.

Synthetic kitchen sponges are made from cellulose. During the processing of the sponges, tiny bubbles of air become trapped in the cellulose.

---

layers of cells with a layer of jellylike material between them. Wastes pass out the large opening at the top of the sponge.

CONTENT DEVELOPMENT

Emphasize that even though sponge cells are capable of functioning on their own, there is nevertheless some specialization, or division of labor, among the cells. The cells making up the outer layer of the sponge's body are flat, and they serve as a protective covering. The cells

of the inner layer contain flagella, which set up water currents that bring in food particles. These cells, called collar cells, capture and digest food particles that may be in the water.

In the jellylike mass between these two cell layers are specialized cells called amoebocytes. These cells secrete skeletal materials and transport digested food. Some amoebocytes can also form sperm and egg cells that function in sexual reproduction.

• **How do these animals take in food and oxygen?** (As ocean water moves into a sponge through its pores, the inner cells remove food and oxygen.)

• **Why is the name *Porifera* used for the sponge phylum?** (A sponge's body contains numerous pores.)

● ● ● ● **Integration** ● ● ● ●

Use the discussion of the derivation of the word *porifera* to integrate concepts of language arts into your lesson.

1-4 Cnidarians

MULTICULTURAL
OPPORTUNITY 1-4

Tell students that even though reef islands, especially those close to sea level, are not very stable, many people live on coral reefs. Have students research the South Pacific civilizations located in the Coral Seas. How do the people live in harmony with their environment, obtaining their food from the waters surrounding them?

ESL STRATEGY 1-4

Help students complete the following chart illustrating the differences between sponges and cnidarians. Students should write yes or no in each box in the chart. The answers are shown in parentheses.

Yes or No?	Sponges	Cnidarians
Backbones?	(No.)	(No.)
Specialized tissues?	(No.)	(Yes.)
Reproduce sexually?	(Yes.)	(Yes.)
Reproduce asexually?	(Yes.)	(Yes.)
Single opening?	(No.)	(Yes.)

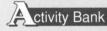ctivity Bank

An Unusual Cnidarian, Activity Book, p. 137. This activity can be used for ESL and/or Cooperative Learning.

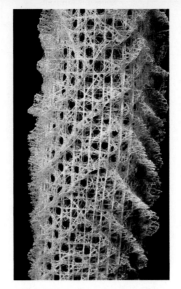

Figure 1-11 *The lacy skeleton of this glass sponge consists of thousands of spicules made of glassy material. What is the function of spicules?* ❶

(male cells). These cells join, and a young sponge develops. **Asexual reproduction** is the process by which a single organism produces a new organism. Sponges reproduce asexually by budding. In budding, part of a sponge simply falls off the parent sponge and begins to grow into a new sponge. During cold winters, some freshwater sponges produce structures that contain groups of cells surrounded by a hard, protective layer. When conditions become favorable again, these structures grow into new sponges.

In addition to being the source of natural sponges, sponges are also an important source of powerful antibiotics that can be used to fight disease-causing bacteria and fungi. Sponges also provide homes and food for certain worms, shrimps, and starfishes.

1-3 Section Review

1. What are the main characteristics of sponges?
2. How do food and oxygen enter a sponge's body?
3. How do sponges reproduce?

Critical Thinking—*Making Predictions*
4. Predict what would happen to a sponge if it lived in water that contained a great deal of floating matter.

Guide for Reading

Focus on this question as you read.

▶ What are the main characteristics of cnidarians?

1-4 Cnidarians

The phylum Cnidaria (nigh-DAIR-ee-uh) consists of many invertebrate animals with dazzling colors and strange shapes. The animals that make up this phylum include corals, jellyfishes, hydras, and sea anemones. As you can see in Figure 1-12, cnidarians have two basic body forms: a vase-shaped polyp (PAHL-ihp) and a bowl-shaped medusa (muh-DOO-suh). A polyp usually remains in one place; a medusa can move from place to place.

Cnidarians have a hollow central cavity with only one opening called the mouth. The phylum

---

1-3 (continued)

INDEPENDENT PRACTICE

Section Review 1-3
1. Sponges have pores. They show little or no movement and can reproduce both sexually and asexually.
2. Water carries food and oxygen through the pores in the sponge.
3. Either sexually by the joining of a male and a female cell or asexually by budding.
4. The pores of the sponge might be-

come clogged, preventing the passage of food and water.

REINFORCEMENT/RETEACHING

Review students' responses to the Section Review questions. Reteach any material that is still unclear, based on students' responses.

CLOSURE

▶ *Review and Reinforcement Guide*
Have students complete Section 1-3 in the *Review and Reinforcement Guide.*

TEACHING STRATEGY 1-4

FOCUS/MOTIVATION

Introduce this section by showing the class a simple model of a jellyfish. A plastic bowl, such as a soft-margarine container, could be used to represent the body. With the dish turned upside down, tape some pieces of string or yarn around the rim to resemble the tentacles.

Explain that the animals of the next

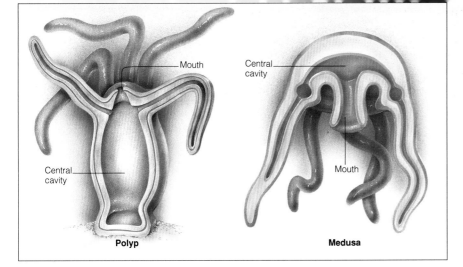

Figure 1–12 *Cnidarians have two basic body forms: a vase-shaped polyp and a bowl-shaped medusa. What is the difference between a polyp and a medusa?* ②

① name Cnidaria is taken from the Greek word meaning to sting. All cnidarians have **nematocysts,** which are special stinging structures. Nematocysts are used to stun or kill a cnidarian's prey. It is no surprise then that nematocysts are found on the tentacles surrounding a cnidarian's mouth. After capturing and stunning its prey, a cnidarian pulls the prey into its mouth with the tentacles. Once the food is digested within its central cavity, a cnidarian releases waste products through its only opening, which is its mouth.

Unlike sponges, cnidarians contain groups of cells that perform special functions. In other words, cnidarians have specialized tissues. An example of a specialized tissue is a nerve net, which is a simple nervous system that is concentrated around a cnidarian's mouth but spreads throughout the body. An interesting characteristic of cnidarians can be seen in Figure 1–13. Notice that a cnidarian is symmetrical. If you drew a line through the center of its body, both sides would be the same.

Most cnidarians can reproduce both sexually and asexually. Like sponges, cnidarians reproduce asexually by budding and sexually by producing eggs and sperm.

Figure 1–13 *Jellyfishes have a type of symmetry in which their body parts repeat around an imaginary line drawn through the center of their body. What other invertebrates have this type of symmetry?* ③

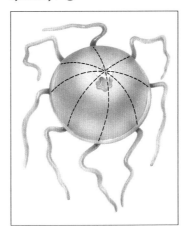

C ■ 21

to look up the meaning of the word *cnidaria* in the textbook. Ask another student to read the definition of *nematocyst*. Point out that the presence of nematocysts is a trait unique to cnidarians.

Have students examine Figure 1–12.

• **Do you see anything in these body forms that is different from the body form of a sponge?** (The most important difference students should notice are the two structures that the sponge does not have: tentacles and a mouth opening.)

Explain that cnidarians are the simplest animals that have a distinctive gut in which digestion takes place. Remind students that there is no specific gut area in sponges—digestion takes place in the individual cells.

● ● ● ● **Integration** ● ● ● ●

Use the discussion of the derivation of the word *cnidaria* to integrate concepts of language arts into your lesson.

ENRICHMENT

You may wish to point out that one measure of increasing complexity among animal organisms is the type of gut. The simplest animals, the sponges, have no distinct gut. The next-simplest animals have a gut with only one opening, the mouth. On the next level of complexity, animals have what is often referred to as a complete gut, with both a mouth and an anus. In the more complex organisms, the gut will contain organs and will eventually be part of an organ system.

group to be studied all have a hollow body cavity. Point to the model of the jellyfish.

• **How many openings lead into the body cavity in this group of animals?** (Students should indicate that there is one opening.)

• **What do these pieces of string represent?** (Tentacles.)

• **Where are the tentacles located?** (The tentacles form a ring about the mouth.)

• **What do you think is the function of these tentacles?** (Some students may be

aware of the nematocysts, or stinging structures, of the tentacles that function in capturing prey.)

• **What kind of animal does this model represent?** (Lead students to answer that the model represents a jellyfish, but explain that a jellyfish is only one kind of cnidarian.)

CONTENT DEVELOPMENT

Write "cnidarians" and "nematocysts" on the chalkboard. Ask a student

Figure 1–14 *Unlike most cnidarians, which live in salty oceans, these green hydras live in freshwater lakes and streams. How do hydras reproduce?* ❶

Hydras

Hydras belong to a group of cnidarians that spend their lives as polyps (vase-shaped body forms). Hydras are the only cnidarians that live in fresh water. All other cnidarians live in salty oceans. Unlike most other types of polyps, hydras can move around with a somersaulting movement. They can reproduce either asexually by budding or sexually by producing eggs and sperm.

Corals

Like all cnidarians, corals are soft-bodied organisms. However, corals use minerals in the water to build hard, protective coverings of limestone. When a coral dies, the hard outer covering is left behind. Year after year, for many millions of years, generations of corals have lived and died, each generation adding a layer of limestone. In time, a coral reef forms. The Great Barrier Reef off the coast of Australia is an example of a coral reef. The top layer of the reef contains living corals. But underneath this "living stone" are the remains of corals that may have lived millions of years ago when dinosaurs walked the Earth. ❶

Corals are polyps that live together in colonies. These colonies have a wide variety of shapes and

Figure 1–15 *Corals—such as staghorn corals (top left), lettuce corals (top right), grooved brain corals (bottom left), and tube corals (bottom right)—produce skeletons of calcium carbonate, or limestone.*

22 ■ C

ing on their basal disks. They can also form an air bubble at their bases and float upside down.

colors. Some coral colonies look like antlers. Others resemble fans swaying in the water. Still others resemble the structure of a human brain.

At first glance, a coral appears to be little more than a mouth surrounded by stinging tentacles. See Figure 1–15. But there is more to a coral than meets the eye. Living inside a coral's body are simple plant-like autotrophs known as algae. Algae help to make food for the coral. But because algae need sunlight to make food, corals must live in shallow water, where sunlight can reach them. The relationship between a coral animal and an alga plant is among the most unusual in nature. **②**

CONNECTIONS

No Bones About It ❸

What do coral and bone have in common? Until a few years ago, it was thought they had little in common. Today, however, there are more similarities between coral and bone than meet the eye. And these similarities have made coral an excellent stand-in for bones in repairing serious bone fractures, or breaks.

Because coral resembles bone, doctors have recently begun using it to replace bone. Before the coral is inserted into the body, however, it is heated. The heat changes limestone (calcium carbonate), which is the main substance in coral, into calcium-containing hydroxyapatite, which is the main ingredient in bone. In the process, living coral organisms are killed.

When the coral is placed at the site of the fracture, it blends almost seamlessly with the bone. The maze of channels within the coral provides passageways through which blood vessels from nearby bone can grow. A permanent bond between bone and coral results.

Over time, the coral becomes filled with new bone.

Because coral is made of material naturally found in bones, the body does not treat it as a foreign substance. In other words, coral does not seem to activate the body's immune system (the body's defense against foreign material) nor produce inflammation (a condition that causes redness, pain, and swelling). Thus the coral remains unaffected, and the body remains unharmed as new bone grows.

BACKGROUND INFORMATION
COELENTERATA

Phylum Cnidaria has in the past been known as phylum Coelenterata. The word *coelenterata* means "hollow intestine."

1–4 (continued)

CONTENT DEVELOPMENT

Tell the class to look at Figure 1–16 and then read the textbook section on sea anemones.

• **How did you distinguish the clownfish from the tentacles of the sea anemone?** (Accept all answers. Students will likely mention that the clownfish is orange and white, whereas the tentacles of the anemone are a different color.)

• **In what way is the relationship between the clownfish and sea anemone similar to that of the algae inside the coral's body?** (Students should point out they benefit each other. The relationship is symbiotic.)

• **How do the clownfish and anemone benefit each other?** (The anemone protects the clownfish from predators, and the clownfish attracts other food fishes to the anemone.)

GUIDED PRACTICE

Skills Development

Skill: Making comparisons

Ask the class to look at the sea anemone in Figure 1–16 and the jellyfish in Figure 1–17. Point out that even though both of these organisms are cnidarians, their body plans are different.

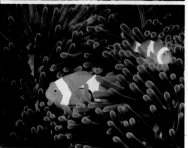

Figure 1–16 *The nematocysts on the tentacles of sea anemones are used to catch food. Although most large sea anemones eat fishes, this clownfish swims undisturbed through the sea anemone's tentacles. How do these two organisms help each other?* **1**

Figure 1–17 *Jellyfishes, such as Aequorea, spend most of their life as bowl-shaped medusas.*

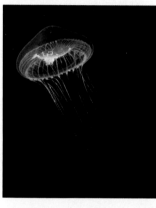

24 ■ C

Sea Anemones

Can you see the fish swimming through what looks like a plant in Figure 1–16? That plant is actually an animal—a cnidarian known as a sea anemone (uh-NEHM-uh-nee). Sea anemones are polyps that resemble underwater flowers. Their "petals," however, are really tentacles that contain nematocysts (stinging structures). When a fish passes near a sea anemone's tentacles, the nematocysts stun the fish. Then the tentacles pull the fish into the sea anemone, where the stunned prey is digested within the central cavity.

If a fish is lucky enough to be a clownfish, however, it can swim unharmed through a sea anemone's tentacles. This friendly relationship between a sea anemone and a clownfish protects a clownfish from other fishes that might try to attack it. At the same time, a clownfish serves as living bait for a sea anemone. When other fishes see a clownfish swimming among a sea anemone's tentacles, they swim nearer, hoping to make the clownfish their next meal. But before they know it, they are grabbed by the sea anemone's tentacles and become its next meal!

Jellyfishes

If you have ever seen a jellylike cup floating in the ocean near you, you probably knew enough to stay clear of it. This cnidarian, known as a jellyfish, is one that most people recognize immediately.

Although a jellyfish may look harmless, it can deliver a painful poison through its nematocysts, which are located on its tentacles. In fact, even when the nematocysts are broken up into small pieces, they remain active. They can sting a passing swimmer who accidentally bumps into them. The largest jellyfish ever found had tentacles that reached out for more than 30 meters.

Of course, the nematocysts are not there merely to disturb unsuspecting swimmers. Like all cnidarians, jellyfishes use the nematocysts to capture prey. One type of jellyfish, the sea wasp jellyfish, produces a strong nerve poison that has helped scientists to better understand the function of nerves in humans.

• **What differences do you see in the shapes of a jellyfish and a sea anemone?** (Lead students to suggest that a jellyfish has a bell-shaped body, and a sea anemone has a cup or tubelike form.)

• **Do both of these cnidarians have tentacles around the mouth?** (Yes.)

• **In jellyfish, do the mouth and tentacles point upward or do they point downward?** (They point downward.)

• **In what direction do the tentacles and mouth point in a sea anemone?** (They point upward in a sea anemone.)

Remind students that all cnidarians have a body plan like that of either the jellyfish (medusa) or the sea anemone (polyp).

• **Which of these body forms is stationary and which is adapted for swimming around?** (The polyp is usually attached to an underwater surface, whereas the medusa is adapted for swimming.)

1-4 Section Review

1. What is the main characteristic of cnidarians?
2. What is the function of nematocysts?
3. What is a polyp? A medusa?
4. What animals make up the phylum Cnidaria?

Critical Thinking—*Making Comparisons*

5. Which phylum is more complex—the sponges or the cnidarians? Explain your answer.

1-5 Worms

Most people think of worms as slimy, squiggly creatures. And, in fact, many are. There are, however, many kinds of worms that look nothing like the worms used to bait fishing hooks. You can see examples of such worms in Figure 1–18. Worms are classified into three main phyla based on their shapes. **The three phyla of worms are flatworms, roundworms, and segmented worms.**

If you were to draw an imaginary line down the entire length of a worm's body, you would discover that the right half is almost a mirror image of the left half. See Figure 1–19 on page 26. With the exception of one phylum, this body symmetry

Guide for Reading

Focus on these questions as you read.

▶ What are the three main groups of worms?
▶ What are the characteristics of each group of worms?

Figure 1-18 *Contrary to popular belief, many worms—such as the feather duster worm (left) and the sea mouse (right)—are neither squiggly nor are they slimy.*

C ■ 25

1-5 Worms

MULTICULTURAL OPPORTUNITY 1-5

Have students investigate ways in which worms are helpful, such as their use in controlling insect species and their use by organic gardeners in fertilizing and aerating gardens. Can students find examples from different areas of the world?

ESL STRATEGY 1-5

Dictate the following questions and ask ESL students to work with English-speaking partners in preparing their answers.

1. What are the three main phyla of worms? (Flatworms, roundworms, and segmented worms.)
2. Which phylum can regrow missing parts? What is this process called? (Some of the flatworms; regeneration.)
3. Which phylum has a well-developed digestive system (compared with that of cnidarians), can live on land, in water, in plant tissue, or in a human's body? (Roundworms.)

Activity Bank

Flat as a Worm, Activity Book, p. 141. This activity can be used for ESL and/or Cooperative Learning.

TEACHING STRATEGY 1-5

FOCUS/MOTIVATION

Have students look at the main headings in Section 1–5. Then hold up a piece of ribbon, a short piece of rope or string, and a single strand of some beads on a string.

• **Which of these is similar to those worms called flatworms?** (The flatworm's body is like a ribbon.)
• **Which of the other two groups of worms is most like the piece of rope?** (Because roundworms have a rounded body, they are somewhat like the rope.)
• **How are segmented (annelid) worms like this string of beads?** (The bodies of segmented worms are made up of sections like beads on a string.)

INDEPENDENT PRACTICE

Section Review 1-4

1. Cnidarians have a hollow central body cavity with only one opening. They also have stinging cells called nematocysts.
2. To sting and kill prey.
3. A polyp is a vase-shaped body form; a medusa is a bowl-shaped form.
4. Corals, jellyfishes, hydras, and sea anemones.
5. Cnidarians are more complex because they have specialized tissues.

REINFORCEMENT/RETEACHING

Review students' responses to the Section Review questions. Reteach any material that is still unclear, based on students' responses.

CLOSURE

▶ *Review and Reinforcement Guide*

Have students complete Section 1–4 in the *Review and Reinforcement Guide*.

TAPEWORMS

The largest tapeworm that affects humans is the broad fish tapeworm. This parasite can grow to be about 18 meters long with a body composed of up to 4000 sections. In addition to humans, this worm may also infect bears, dogs, foxes, and cats. Among humans, it is most common in certain parts of the Baltic region, where almost the entire population is infected.

BACKGROUND INFORMATION

NEMATODES

The nematodes are second only to the insects in the number of species. About 10,000 species are known, but it is estimated that more than 500,000 species may exist. Virtually every species of vertebrate animal harbors nematode parasites. Additionally, others are parasites of plants, but the most numerous of all are the free-living forms. Free-living nematodes are common in soil, mud, and on rotting organic matter. They are also found in fresh water and in every ocean. Roundworms play an important role in decomposing organic matter in both terrestrial and aquatic environments.

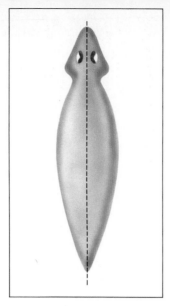

Figure 1–19 *Most of the more complex animals, such as the worms, have bodies in which the left half and the right half are almost mirror images of each other.*

Figure 1–20 *The planarian (right) is an example of a flatworm that is able to regenerate. An injury to this planarian (left) divided its head in half, and then the two halves regenerated their lost parts.*

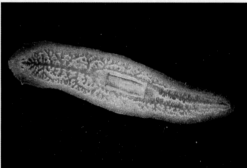

26 ■ C

(sameness of form on either side of a dividing line) is characteristic of all animals you will read about from here on, including worms.

Flatworms

Flatworms, as you might expect, have flat bodies. They are grouped in the phylum Platyhelminthes (plat-ih-hehl-MIHN-theez). The word platyhelminthes comes from two Greek words, *platy-*, meaning flat, and *helminth,* meaning worm. Some scientists believe that flatworms evolved from cnidarians. Other scientists are convinced that flatworms developed independently.

You have probably never seen a flatworm, much less a flatworm like the one in Figure 1–20. This flatworm is called a planarian. Most planarians, which are barely 0.5 centimeter long (a little less than the width of the nail on your pinky), live in ponds and streams—often on the underside of plant leaves or on underwater rocks. Planarians feed on dead plant or animal matter.

When there is little food available, however, some planarians do a rather unusual thing. They digest their own body parts. Later, when food is available once again, the missing parts regrow. Interestingly, if a planarian is cut into pieces, each piece eventually grows into a new planarian! The ability of an organism to regrow lost parts is called **regeneration.**

Some flatworms grow on or in living things and are called **parasites.** Tapeworms, which look like long, flat ribbons, live in the bodies of many animals, including humans. The head of a tapeworm has

---

CONTENT DEVELOPMENT

Have students examine Figure 1–19 and read the caption.
• **What characteristic makes the more complex worms different from cnidarians?** (Bilateral body symmetry.)

Explain that the first type of worm discussed, the flatworm, is called an unsegmented worm. The flatworm is more complex than the cnidarians because it has three layers of cells rather than two. Flatworms belong to the phylum Platyhelminthes.

● ● ● ●   **Integration**   ● ● ● ●

Use the discussion of the word *platyhelminthes* to integrate concepts of language arts into your lesson.

ENRICHMENT

Take several specimens of live planarians and cut them into three pieces. This can be done by placing each planarian on a flat microscope slide, holding the slide over an ice cube to chill the worm, then using a scalpel or razor blade to cut the worm. Put each section in a separate petri dish labeled head, middle, tail. Fill each petri dish half full with aquarium water; cover the dishes and place them in a dark place. Have students observe the

sections three times a week for three weeks and record what they see.

CONTENT DEVELOPMENT

Explain that the parasitic tapeworms are also members of the phylum Platyhelminthes. Tell the class that humans can become infected by tapeworms by eating contaminated beef that is not cooked well.

Briefly explain the life cycle of the beef tapeworm.

special hooks that are used to attach it to the tissues of a **host**, or the organism in which it lives. The tapeworm causes illness by taking a host's food and water, as well as by producing wastes and blocking the host's intestines.

A tapeworm can grow as long as 6 meters inside its host. However, size is not always a good indicator of danger. The most dangerous human tapeworm is only about 8 millimeters long and enters the body through microscopic eggs in some types of food.

Roundworms

You probably have been told never to eat pork unless it is well cooked. Do you know the reason for this warning? It has to do with a type of roundworm called *Trichinella* (trih-KIGH-nehl-uh), which lives in the muscle tissue of pigs. If a person eats a piece of raw or undercooked pork, *Trichinella* that are still alive in the pork enter the person's body. As many as 3000 roundworms can be contained in a single gram of raw or undercooked pork!

Once inside the body, the roundworms live and reproduce in the intestines of the host. Female roundworms release hundreds of immature roundworms, which are carried in the bloodstream. These immature roundworms then burrow into surrounding tissues and organs. This causes terrible pain for the host. Once inside tissues and organs, *Trichinella* become inactive. The name of this disease is trichinosis (trihk-ih-NOH-sihs).

Roundworms resemble strands of spaghetti with pointed ends. They belong to the phylum Nematoda (nehm-uh-TOHD-uh). The word nematode means threadlike. Roundworms live on land or in water. Many roundworms are animal parasites, although some live on plants. One type of roundworm, the hookworm, infects more than 600 million people in the world every year. Hookworms enter the body by burrowing through the skin on the soles of the feet. They eventually end up in the intestines of their hosts, where they live on the blood.

Figure 1–21 *Some flatworms, such as the tapeworm, are parasites. Notice that the head of a tapeworm has suckers and other structures that enable it to attach to the inside of its host's intestines. What is a parasite?* ❶

Figure 1–22 Trichinella *worms, which cause the disease trichinosis, burrow into the muscle tissue of their host (top). These threadworms, tunneling through the tissues of a sheep's small intestine, are parasitic roundworms (bottom).*

cycle? (Cooking meat thoroughly before it is eaten.)

● ● ● ● **Integration** ● ● ● ●

Use the discussion of parasites to integrate concepts of symbiosis into your lesson.

Use the discussion of trichinosis to integrate concepts of food service into your lesson.

GUIDED PRACTICE

▶ *Laboratory Manual*

Skills Development

Skills: Making observations, making comparisons, making inferences

At this point you may want to have students complete the Chapter 1 Laboratory Investigation in the *Laboratory Manual* called Flatworms and Roundworms. In the investigation students will compare the characteristics of flatworms and roundworms.

CONTENT DEVELOPMENT

Emphasize that the roundworms are the first group of animals to be studied that have a complete digestive system. Roundworms, like flatworms, are unsegmented worms.

● **In what important way is the digestive system of roundworms different from that of cnidarians and flatworms?** (Roundworms have a complete digestive system with two openings—a mouth at the head end and an anus at the tail end.)

1. A cow eats grass contaminated with tapeworm eggs.
2. The eggs pass to the cow's intestines, where they develop into immature worms.
3. The worms are carried by the cow's blood to its muscles.
4. Poorly cooked meat (muscles) is eaten by a human. The immature worms develop into mature tapeworms in the human's intestine.
5. The mature worms produce eggs,

which are passed out of the human in solid wastes.
6. The eggs cling to grass and soil.
● **How does the life cycle of the beef tapeworm point up the need for effective sewage treatment?** (Because tapeworm eggs are eliminated in the feces of an infected human, proper treatment of sewage could destroy the eggs before they reach the grass or soil. This would break the cycle.)
● **What other precaution will break the**

ANNELIDS AND EVOLUTION

Annelids give us many clues about major evolutionary trends. Their ancestors, who were segmented, foreshadowed a trend toward increased size and greater complexity of internal organs. Annelids have a rudimentary brain and segmented ganglia, which means that they have the beginnings of a centralized nervous system with regional neural connections. They have a complete digestive system with regional specializations; they also have a closed circulatory system linked to nephridia, which are the annelid equivalent of vertebrate kidneys.

INTEGRATION

LANGUAGE ARTS

Many slang and colloquial expressions in the English language refer to worms. For example, "the worm in the apple" is often used to describe the one bad thing in an otherwise good situation or idea. To "worm your way out of something" means to wiggle out of unpleasant circumstances, usually by using devious or covert means.

1–5 (continued)

FOCUS/MOTIVATION

Display a map of the world and point to the island of Madagascar (off the southeastern coast of Africa) and to the continent of Antarctica.

• **What kind of animal lives everywhere in the world except for these two places?** (Answers may vary.)

CONTENT DEVELOPMENT

Tell students that annelids or segmented worms, live everywhere in the world except in Madagascar and Antarctica. Annelids live in fresh water, ocean water, and on land. Point out that the most familiar representative of this phylum is the earthworm but that there are actually about 9000 other species of segmented worms.

• **What is one obvious difference between earthworms and roundworms?**

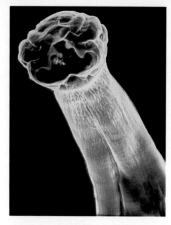

Figure 1–23 *Hookworms use the sharp teeth and hooks on their head end to burrow through their host's skin.*

Figure 1–24 *Many segmented worms, such as the sandworm, use hooklike jaws to capture their prey. To which phylum of invertebrates do sandworms belong?* ❶

28 ■ C

Like all worms, roundworms have a head and a tail. In fact, worms were the first organisms to evolve with distinct head and tail ends. Roundworms have a tubelike digestive system that has two openings—a mouth and an anus (AY-nuhs). Food enters the digestive system through the mouth (in the head end) and waste products leave the digestive system through the anus (in the tail end). Although the digestive system of roundworms may not seem complex, it is far more advanced than that of cnidarians. As you may recall, cnidarians have one opening for both the intake of food and the elimination of wastes.

Segmented Worms

Segmented worms belong to the phylum Annelida (an-uh-LIHD-uh). The term annelid comes from the Latin word for ringed. This is an appropriate name for these invertebrates because their most obvious feature is a ringed, or segmented, body. These segments are visible on the outside of an annelid's body. Segmented worms live in salty oceans, in freshwater lakes and streams, or in the soil.

The segmented worm you are probably most familiar with is the common earthworm. The body of an earthworm is divided into numerous segments—100 at least. If you have ever touched an earthworm, you know it has a slimy outer layer. This layer, made of a slippery substance called mucus, helps the earthworm to glide through the soil. Tiny setae (SEE-tee; singular: seta), or bristles, on the segments of the earthworm also help it to pull itself along the ground.

As all gardeners know, earthworms are essential to a healthy garden. By burrowing through the soil, earthworms create tiny passageways through which air enters, thus improving the quality of the soil. As earthworms burrow, they feed on dead plant and animal matter. These materials are only partially digested. The undigested portions, or waste products, are eliminated into the surrounding soil, thus fertilizing the soil. The digestive system of an earthworm is shown in Figure 1–25.

Earthworms have a closed circulatory system. A closed circulatory system is one in which all body fluids are contained within small tubes. In an

(Earthworms are segmented; roundworms are not.)

Explain that segmented worms are the most complex worms. Like the unsegmented worms, annelids are bilaterally symmetrical. Their body plan, however, is more complicated than that of the unsegmented worm. The body of an annelid is divided into many small segments, most of which appear to be almost identical.

An important characteristic of anne-

lids is that they have a body cavity, or coelom, filled with fluid. This gives segmented worms a body structure that is like a "tube within a tube."

GUIDED PRACTICE

Skills Development

Skills: Making observations, making comparisons, making inferences

Have students complete the in-text Chapter 1 Laboratory Investigation: Observing Earthworm Responses. In the

earthworm, the fluids are pumped throughout its body by a series of ringed blood vessels found near its head region. Because these blood vessels help to pump fluids through the circulatory system, they are sometimes called hearts.

Earthworms, like most annelids, have no special respiratory organs. Oxygen enters through the skin, and carbon dioxide leaves through the skin. In order for this to happen, the skin must stay moist. If an earthworm's skin dries out—which might happen if the earthworm remains out in the heat of the sun—the earthworm suffocates and dies.

Although earthworms have only a simple nervous system, they are very sensitive to their environment. An earthworm's nervous system consists of a brain found in the head region, two nerves that pass around the intestine, and a nerve cord located on the lower side.

In addition to these structures, an earthworm has a variety of cells that sense changes in its environment. One group of such cells, located on the first few body segments, detects moisture. Recall how important moisture is to an earthworm's survival. If an earthworm emerges from its burrow and encounters a dry spot, it will move from side to side until it finds dampness. If it does not find any moisture, it will retreat into its burrow. An earthworm can also sense danger and warn other earthworms. It does so by releasing a sweatlike material that helps it to glide more easily and to warn others of the nearby danger.

Figure 1–25 *The drawing shows the digestive system of the earthworm, a segmented worm. The photograph is of an earthworm moving along the surface of the soil. What structures help the earthworm move?* ❷

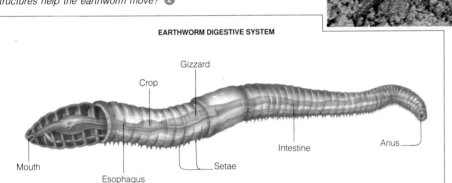

EARTHWORM DIGESTIVE SYSTEM

Gizzard
Crop
Mouth
Esophagus
Setae
Intestine
Anus

ACTIVITY
DISCOVERING

Worms at Work

1. Put a tall, thin can with the closed end up in a large jar. Fill the large jar with soil to the level of the tin can. Add a thin layer of sand.

2. Put five earthworms in the jar. Add pond water or tap water that has been standing for a day to slightly moisten the soil. Add more water if the soil appears dry.

3. Cover the outside of the jar with black paper and leave the jar undisturbed for a day.

4. Uncover the jar and look for the earthworms. Record your observations.

5. Cover the jar for one more day. Repeat step 4.

■ Have the earthworms been at work? Explain.

ACTIVITY
DISCOVERING
WORMS AT WORK

Discovery Learning

Skills: Applying concepts, making observations, recording data, making inferences

Materials: tall can, large jar, soil and sand, 5 earthworms, pond water, black paper

In observing an earthworm, students should be able to tell the difference between the head and the tail by noticing that the head end is located closer to the clitellum than the tail ends is. The clitellum is the light-colored band around the earthworm. Other distinguishing features are a lip at the head end extending over the mouth and a tail that is usually more pointed than the head. Students should notice that the earthworms can move forward or backward by using their setae, or bristles. The earthworms will be busy burrowing through the soil.

• **What happens to undigested food?** (It passes through the intestine and out of the worm through the anus.)

INDEPENDENT PRACTICE

▶ *Activity Book*

Students who need practice with the concepts of this section should be provided with the Chapter 1 activity called Comparing Digestive Systems of Worms.

REINFORCEMENT/RETEACHING

To help students compare the three groups of worms, have them prepare a chart with these headings: Group of Worms, Structure, Examples.

---

investigation students will explore how earthworms respond to changes in moisture and light.

CONTENT DEVELOPMENT

Emphasize that compared with other worms, segmented worms have a more complete digestive system, a better developed circulatory system, and a more highly developed nervous system. Have students examine the diagram of the earthworm's digestive system in Figure 1–25.

• **If the digestive tract begins with the mouth and ends with the anus, how much of the earthworm's body is taken up by the digestive tract?** (The digestive tract runs the entire length of the worm's body.)

• **Through which organs must food pass before it reaches the intestine?** (Mouth, esophagus, crop, and gizzard.)

• **Where is the intestine located in the digestive tract?** (Before the anus, in the posterior portion of the worm.)

1-6 Mollusks

MULTICULTURAL OPPORTUNITY 1-6

In Japan, a mollusk called the *awabi* (also called the western abalone) has been regarded from ancient times as a valuable gift. Its dried and sliced meat is wrapped in red and white papers and given to friends as a symbol of joy and happiness. Another sea product regarded as a valuable gift by the Japanese is a fish called *katsuo*. After being cut into strips, which are dried in the sun until they become as hard as wood, it is placed on a lacquered tray or in an attractive bamboo basket and given as a token of congratulations to celebrate success of any kind.

Ask students to share information about customs involving gifts that are part of their own cultural heritage. What are some natural products that are traditionally given on festive occasions?

ESL STRATEGY 1-6

Have students prepare labeled illustrations (models or drawings) of different types from each of the three phyla of common mollusks. Furnish reference materials to ensure proper labeling.

1-5 (continued)

GUIDED PRACTICE

▶ *Laboratory Manual*

Skills Development

Skills: Making observations, interpreting diagrams, making inferences

At this point you may want to have students complete the Chapter 1 Laboratory Investigation in the *Laboratory Manual* called The Earthworm. In the investigation students will explore ways in which the earthworm is adapted to its environment.

INDEPENDENT PRACTICE

Section Review 1-5

1. The three phyla discussed in the text are flatworms, roundworms, and segmented worms. Examples, respectively,

---

Figure 1-26 *An earthworm contains both male and female structures. When two earthworms mate, they exchange only sperm cells.*

Guide for Reading

Focus on these questions as you read.
▶ *What are the main characteristics of mollusks?*
▶ *What are the three main groups of mollusks?*

ACTIVITY DOING

Mollusks in the Supermarket

Visit your neighborhood supermarket and find the mollusks that are being sold as food. Make a chart in which you include the name of the mollusk and the part being sold as food.

30 ■ C

---

1. List the three phyla of worms and give an example of each.
2. Name two types of worms that cause disease in humans.
3. How does an earthworm contribute to a healthy garden?
4. In what ways are worms more complex than cnidarians?

Connection—*Medicine*

5. Imagine that you are a doctor. Recently a number of your patients have developed hookworm infections. Describe some actions you would prescribe to prevent others from being infected by hookworms.

1-6 Mollusks

If you enjoy eating a variety of seafood, you are probably already familiar with some members of the phylum Mollusca (muh-LUHS-kuh). Members of this phylum are called mollusks (MAHL-uhsks). Mollusks include clams, oysters, mussels, octopuses, and squids. The word mollusk comes from the Latin word meaning soft. **Mollusks are soft-bodied animals that typically have inner or outer shells.**

Most mollusks also have a thick muscular foot. Some mollusks use this foot to open and close their shell. Other mollusks use it for movement. Still others use the foot to bury themselves in the sand or mud.

The head region of a mollusk generally contains a mouth and sense organs such as eyes. The rest of a mollusk's body contains various organs that are involved in life processes such as reproduction, circulation, excretion, and digestion. A soft mantle covers much of a mollusk's body. The mantle produces the material that makes up the hard shell. As the mollusk grows, the mantle enlarges the shell, providing more room for its occupant.

The more common mollusks are divided into three main groups based on certain characteristics.

---

are planarians and tapeworms; *Trichinella* and hookworms; earthworms and sandworms.
2. Possible answers include tapeworms, *Trichinella*, and hookworms.
3. It creates passages for air to enter the soil and releases fertilizing materials.
4. Worms have more complex digestive, circulatory, and excretory systems.
5. Possible answers include advising patients to wear shoes and improving sewage treatment to protect water supplies.

REINFORCEMENT/RETEACHING

Review students' responses to the Section Review questions. Reteach any material that is still unclear, based on students' responses.

CLOSURE

▶ *Review and Reinforcement Guide*
Have students complete Section 1-5 in the *Review and Reinforcement Guide*.

These characteristics include the presence of a shell, the type of shell, and the type of foot. **The three main groups of common mollusks are snails, slugs, and their relatives; two-shelled mollusks; and tentacled mollusks.**

Snails, Slugs, and Their Relatives

The largest group of common mollusks are those that have a single shell or no shell at all. The members of this group include snails, slugs, and sea butterflies, to name a few. These mollusks are also known as gastropods (GAS-troh-pahdz). The word gastropod means stomach foot. This name is appropriate because most gastropods move by means of a foot found on the same side of the body as their stomach. Gastropods live in both fresh water and salt water, as well as on land. Those gastropods that live on land, however, still must have a moist environment in order to survive.

Gastropods have an interesting feature called a radula (RAJ-oo-luh) in their mouth. The tongue-shaped radula resembles a file used by carpenters to scrape wood. The radula files off bits of plant matter into small pieces that can easily be swallowed.

The single-shelled gastropod that is probably most familiar to you is a garden snail. Have you ever watched a garden snail as it moves? If so, you might have noticed that it leaves a trail of slippery mucus behind. The mucus enables a snail to slowly glide across different types of surfaces. This is especially helpful when a snail encounters rough surfaces.

Slugs are gastropods that do not have a shell. The absence of a shell may seem to make slugs easy prey for their predators. But slugs are not entirely helpless. Most spend the daylight hours hiding under rocks and logs, thereby staying out of the way of birds and other animals that might eat them. Gastropods such as sea butterflies escape predators by rapidly swimming away. Many sea slugs, or nudibranchs (NOO-dih-brangks), have chemicals in their body that taste bad or are poisonous to unsuspecting predators.

Figure 1–27 *Mollusks—such as garden slugs (top), coquina clams (center), and cuttlefish (bottom)—have soft bodies. Unlike most mollusks, which have internal or external shells, slugs lack shells completely.*

C ■ 31

ACTIVITY
DOING

MOLLUSKS IN THE SUPERMARKET

Skills: Applying concepts, making comparisons

Charts may include clams, oysters, mussels, squids, octopuses, and conches. The soft bodies of bivalves and gastropods are eaten, whereas only the tentacles of octopuses are usually eaten. These mollusks may be found frozen, fresh, or in cans.

to look for differences in the structure and function of the foot as they continue to read about mollusks.

Call on a student to look up and read aloud the derivation of the word *mollusk* in a dictionary. The word is derived from the Latin word *mollis*, meaning "soft."
• **Why is the word *mollusk* appropriate for the animals shown in the figure?** (All mollusks have a soft, fleshy body. Some students may want to talk about the shell that many mollusks possess, but point out that not all mollusks share this trait.)

● ● ● ● **Integration** ● ● ● ●

Use the discussion of the derivation of the word *mollusk* to integrate concepts of language arts into your lesson.

REINFORCEMENT/RETEACHING

▶ *Activity Book*

Students who have difficulty understanding the concepts of this section should be provided with the Chapter 1 activity called Mollusks.

TEACHING STRATEGY 1-6

FOCUS/MOTIVATION

Introduce this section by having students study the photographs in Figure 1–27. Point out that even though these mollusks are quite different, they share some characteristics.

Explain that the soft body of a mollusk is covered by a curtainlike tissue called the mantle. In those mollusks with a limy shell, the mantle secretes that shell. But

be sure to explain that not all shelled animals are mollusks.
• **A turtle has a shell, and so do organisms such as crabs and lobsters. Are these animals also mollusks?** (No, the origin and composition of turtle and crustacean shells are unlike the calcium-carbonate shells of mollusks.)

CONTENT DEVELOPMENT

Emphasize that mollusks are characterized by a muscular foot. Ask students

Sometime in 1986, a cargo ship bound for Ontario emptied some of its ballast water into Lake St. Clair between Lake Huron and Lake Erie. This routine act has been the probable cause of a new environmental problem in all of the Great Lakes.

The cargo ship had begun its journey in European waters. And the ballast water it expelled contained zebra mussels—small brown-striped mussels about the size of a fingernail.

These zebra mussels are native to the Caspian Sea in the Soviet Union and are thriving virtually unchallenged in the Great Lakes. Their large colonies have damaged water supplies and outlet pipes. They pose a threat to other life by stripping water of the algae they eat.

Although coordinated efforts between scientists in the United States and Canada are beginning to result in effective controls for the zebra mussels, their dramatic invasion has emphasized the need for greater controls on the introduction of foreign organisms.

Figure 1-28 Gastropods, the largest group of common mollusks, have a single shell or no shell at all. The single-shelled gastropod that is probably the most familiar to you is the land snail (left); whereas the sea butterfly (right) is not as common a sight. What are some other examples of gastropods? ❶

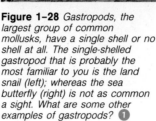

Activity Bank

Moving at a Snail's Pace, p.154

Figure 1-29 Scallops can move around by clapping their shells together (left). A scallop is also able to gather information about its surroundings with the help of its eyes, which are located on the mantle and resemble tiny blue dots (right).

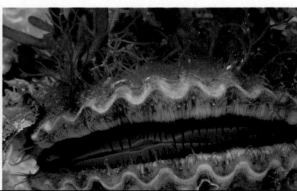

Two-Shelled Mollusks

Clams, oysters, scallops, and mussels are members of the group of common mollusks called bivalves, or two-shelled mollusks. Bivalves have two shells that are held together by powerful muscles. Although most bivalves remain in one place, scallops and a few others can move around rapidly by clapping their shells together. This action forces water out between the shells, thereby propelling the bivalve.

Unlike gastropods, bivalves do not have a radula. Instead, they are filter feeders. This means that as water passes over the body of a bivalve, the bivalve filters out small organisms.

The mantle of a bivalve, like that of most mollusks, contains glands that produce its shell. These glands keep the inside surfaces of the shell smooth. If a foreign object, such as a grain of sand, gets stuck between the mantle and the shell of some types of mollusks, the glands cover it with a shiny secretion. After a few years, the grain of sand becomes completely covered with secretions and is called a pearl.

1-6 (continued)

CONTENT DEVELOPMENT

Ask the class to look at the snail in Figure 1-28. Point out that the "eyes" at the ends of the stalks function largely to detect changes in light intensity. There is also a shorter pair of stalks on the head, and these can be retracted. It is believed that a snail's organs of smell are located here.

ENRICHMENT

Many slugs and snails are considered harmful because they do much damage to the plants on which they feed. Additionally, some snails are dangerous because they serve as intermediate hosts for such animal parasites as the sheep liver fluke and human blood fluke. Encourage interested students to prepare oral or written reports on some of the ways in which univalve mollusks can be either beneficial or harmful to humans.

FOCUS/MOTIVATION

👄 **Media and Technology**

Students can begin to explore organisms and their environment through use of the Interactive Video/CD ROM called Virtual BioPark, which looks at the octopus in its ocean habitat. In subsequent lessons and chapters, students can continue their exploration by examining organisms such as the giant orb weaving spider, rattlesnake, rock dove pigeon, prairie dog, and cheetah.

GUIDED PRACTICE

▶ *Laboratory Manual*

Skills Development

Skills: Making observations, interpreting diagrams, making inferences

At this point you may want to have students complete the Chapter 1 Laboratory Investigation in the *Laboratory Manual* called Examining a Clam. In the investigation students will identify some of the internal structures of a clam.

Tentacled Mollusks

The most highly developed mollusks are the tentacled mollusks. The tentacled mollusks are also called cephalopods (SEHF-uh-loh-pahdz). Included in this group are octopuses, squids, and nautiluses. Most cephalopods do not have an outer shell but have some part of a shell within their body. The one exception is the chambered nautilus. The chambered nautilus got its name from the fact that its shell consists of many chambers. These chambers are small when the animal is young but increase in size as the animal grows. As the nautilus grows, it builds a new outer chamber in which it lives.

All cephalopods have tentacles that they use to move themselves and to capture food. But the number and type of tentacles differ from one kind of cephalopod to another. Octopuses, of course, have eight tentacles (the prefix *octo-* means eight); squids have ten.

Even though most cephalopods do not have an outer shell, they do have ways to protect themselves. Most cephalopods can move quickly by either swimming or crawling. Octopuses, squids, and nautiluses can also move by using a form of jet propulsion. Water is drawn into the mantle cavity and then forced out through a tube, propelling the cephalopod backward—away from the danger. In addition, squids and octopuses produce a dark-colored ink when they are frightened. As this ink is released into the water, it helps to hide the mollusk and confuse its predators. The squid or octopus can then escape.

Figure 1-30 *The nautilus (top), squid (center), and octopus (bottom) are examples of cephalopods. What are some characteristics of cephalopods?* ❷

1-6 Section Review

1. What are the main characteristics of mollusks?
2. How are mollusks grouped?
3. List the three main groups of mollusks and give an example of each.
4. What is a mantle? A radula?

Critical Thinking—*Relating Facts*
5. Suppose you found a mollusk with one shell and eyes located on two stalks sticking out of its head. Into which group of mollusks would you place it? Explain.

C ■ 33

Laboratory Investigation

OBSERVING EARTHWORM RESPONSES

BEFORE THE LAB

1. Gather all materials at least one day prior to the investigation. You should have enough supplies to meet your class needs, assuming six students per group.
2. Be sure to keep the earthworms moist up until the time of the experiment. Use the pond or tap water to slightly moisten the soil in which the earthworms are stored. Add more water whenever the soil appears dry.
3. Make sure all the desk lamps are working. Have on hand a few extra bulbs in case one burns out.

PRE-LAB DISCUSSION

Earthworms are among the most familiar of organisms, yet probably few students have taken the time to observe earthworms closely. In this simple demonstration students can observe the basic physical characteristics of earthworms.

Have students read the complete laboratory procedure. Discuss the procedure by asking questions similar to the following:

• **What is the purpose of this investigation?** (To study how earthworms react to changes in moisture and light.)

• **How do you think moisture might be important to an earthworm?** (Answers may vary. Students should recall from the textbook that earthworms depend on moisture to carry out respiration through their skin.)

• **Do you think light is important to an earthworm?** (Answers may vary. Some students may think, incorrectly, that earthworms need light to "see" or to find food; others may realize, correctly, that earthworms prefer dark places where they are protected from predators.)

After students have had a chance to think about the importance of moisture and light, have them write predictions stating what they think the earthworms' responses will be. These predictions can be written in the form of a hypothesis.

• **How do earthworms move? Do you think they can move very quickly?** (Accept all hypotheses at this point.)

Laboratory Investigation

Observing Earthworm Responses

Problem

How do earthworms respond to changes in their environment?

Materials *(per group)*

2 live earthworms in a storage container
medicine dropper
paper towels
tray
piece of cardboard
desk lamp

Procedure 🜔 ·📏· 🔬

1. Open the storage container and examine the earthworms. Record your observations of their physical characteristics. Fill the medicine dropper with water and use it to give your earthworms a "bath." **Note:** *Make sure you keep your earthworms moist by giving them frequent baths.*

2. Fold a dry paper towel and place it on one side of your tray. Fold a moistened paper towel and place it on the other side of the tray.

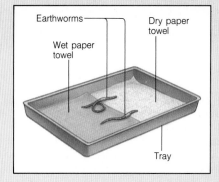

Earthworms — Dry paper towel
Wet paper towel
Tray

3. Place the earthworms in the center of the tray, between the dry paper towel and the moist paper towel. Cover the tray with the piece of cardboard.

4. After 5 minutes, remove the cardboard and observe the location of the earthworms. Record your observations.

5. Return the earthworms to their storage container. Using the dropper, moisten the earthworms with water.

6. Cover the entire bottom of the tray with a moistened paper towel.

7. Place the earthworms in the center of the tray.

8. Cover one half of the tray with the piece of cardboard. Position the lamp above the uncovered side of the tray.

9. After 5 minutes, observe the location of the earthworms. Record your observations.

10. Return the earthworms to their storage container. Using the dropper, moisten the earthworms with water. Cover the container and return it to your teacher.

Observations

Describe the earthworms' color, texture, external features, and other physical characteristics.

Analysis and Conclusions

1. How does an earthworm's response to moisture help it to survive?
2. Does an earthworm's response to light have any protective value? Explain.
3. How is an earthworm's body adapted for movement through the soil?
4. Would you expect to find earthworms in hard soil? Explain.

SAFETY TIPS

Remind students that earthworms, like all living organisms, must be treated with care and in a humane manner.

Remind students of the rules for electrical safety when using the lamps. Emphasize that they should never touch the lamp or cord with wet hands.

TEACHING STRATEGIES

1. As students begin the investigation, have them note the places in the procedure where they are told to moisten the earthworms. Point out that this is essential to the survival of the worms and must be done at least as often as the procedure dictates. Stress that if the worms begin to look dry, students should moisten them immediately.

2. Have teams follow the directions carefully as they work in the laboratory.

Summarizing Key Concepts

1–1 The Five Kingdoms

▲ Today, the most generally accepted classification system contains five kingdoms: monerans, protists, fungi, plants, and animals.

1–2 Introduction to the Animal Kingdom

▲ Vertebrates are animals with a backbone. Invertebrates are animals without a backbone.

1–3 Sponges

▲ Sponges belong to the phylum Porifera. They are called poriferans because their bodies are covered with many pores.

▲ The cells of sponges remove food and oxygen from ocean water as the water flows through pores. The water flowing out through a larger opening carries away waste products.

▲ Sexual reproduction is the process by which a new organism forms from the joining of a female cell (egg) and a male cell (sperm).

▲ Asexual reproduction is the process by which a single organism produces a new organism.

1–4 Cnidarians

▲ Cnidarians belong to the phylum Cnidaria. They have a hollow central cavity with one opening called the mouth.

▲ Cnidarians have structures called nematocysts on the tentacles around their mouth.

1–5 Worms

▲ The three main groups of worms are flatworms, roundworms, and segmented worms.

▲ Flatworms, members of the phylum Platyhelminthes, have flat bodies and live in ponds and streams.

▲ Organisms that grow on or in other living things are called parasites. The organism upon which a parasite lives is called the host.

▲ Roundworms are members of the phylum Nematoda.

▲ Segmented worms, or annelids, have segmented bodies and live in soil or in salt water or fresh water.

1–6 Mollusks

▲ Mollusks, members of the phylum Mollusca, are animals with soft bodies that typically have inner or outer shells.

▲ Most mollusks have a thick, muscular foot and are covered by a mantle.

Reviewing Key Terms

Define each term in a complete sentence.

1–1 The Five Kingdoms
kingdom
autotroph
heterotroph

1–2 Introduction to the Animal Kingdom
vertebrate
invertebrate

1–3 Sponges
spicule
sexual reproduction
asexual reproduction

1–4 Cnidarians
nematocyst

1–5 Worms
regeneration
parasite
host

C ■ 35

DISCOVERY STRATEGIES

Discuss how the investigation relates to the chapter ideas by asking open questions similar to the following:

• **Can you tell the difference between the head and the tail of the earthworm? If so, how?** (Students should be able to see that the head is more rounded and the tail is more pointed, and that there is a lip at the head end extending over the mouth—making observations.)

• **What other features do you notice about the earthworm's body?** (Students should notice the worm's segments and its setae, or bristles—making observations.)

• **What do you notice about the way earthworms move through soil?** (They can move forward or backward by using their setae—making observations.)

OBSERVATIONS

Earthworms will be brown-orange in color and moist and smooth to the touch. The segments of the earthworm will be obvious to students. In addition to identifying the head and tail, they should also notice the light-colored, somewhat swollen band around the earthworm that is closer to the anterior end than to the posterior end. This band is called the clitellum. Student drawings should include the head, tail, segments, and setae.

ANALYSIS AND CONCLUSIONS

1. The skin of the earthworm must stay moist so that oxygen and carbon dioxide can be transmitted. If the earthworm's skin dries out, it suffocates and dies.

2. This response helps the earthworm detect changes in its environment, particularly the approach of predators. A negative response to light would tend to keep the earthworm underground during the day, and this may help it to avoid drying out.

3. The slippery mucus helps the earthworm glide through the soil using its bristlelike setae.

4. No. Earthworms have soft bodies and would be unable to burrow through extremely hard soil.

GOING FURTHER: ENRICHMENT

Part 1

Have students devise an experiment to estimate the speed of earthworms. They may want to investigate whether the earthworms respond more quickly to changes in moisture or to changes in temperature.

Part 2

Provide students with specimens of live planarians for observation. These flatworms can be collected from under rocks or on submerged vegetation in most streams, ponds, and lakes.

Have students use a medicine dropper to place the specimen of planarians on a depression slide. Then have them use a hand lens to examine the planarian and locate the anterior and posterior ends, the ventral and dorsal surfaces, the eyespots, and the sensory lobes.

Next, have students place a small piece of construction paper over half of the hand lens to investigate whether the planarians react to changes in light. Students will find that planarians prefer darkness to light. Remind students that these organisms tend to be found in dark places, such as under stones and logs.

Chapter Review

ALTERNATIVE ASSESSMENT

The *Prentice Hall Science* program includes a variety of testing components and methodologies. Aside from the Chapter Review questions, you may opt to use the Chapter Test or the Computer Test Bank Test in your *Test Book* for assessment of important facts and concepts. In addition, Performance-Based Tests are included in your *Test Book*. These Performance-Based Tests are designed to test science process skills, rather than factual content recall. Since they are not content dependent, Performance-Based Tests can be distributed after students complete a chapter or after they complete the entire textbook.

CONTENT REVIEW

Multiple Choice
1. b
2. d
3. d
4. a
5. d
6. c
7. a
8. b
9. b
10. b

True or False
1. F, heterotrophs
2. T
3. T
4. F, cnidarians
5. F, Roundworms
6. F, mollusks
7. F, two-shelled mollusk
8. F, mollusk

Concept Mapping
Row 1: Monerans, Protists, Fungi, Plants, Animals

CONCEPT MASTERY

1. Accept all logical answers. Students may suggest that the study of animals will help us understand their behavior, protect their environments, or find uses for animals that will benefit people.
2. Monerans: all unicellular without nucleus. Protists: all unicellular with well-defined nuclei. Fungi: similar to plants but do not contain chlorophyll so are not autotrophic. Plants: most multicellular and autotrophic. Animals: all multicellular and heterotrophic.
3. All mollusks have soft bodies. Most have a thick, muscular foot and an inner or outer shell. The gastropods have only one shell or no shell at all. Bivalves have two shells. Cephalopods have tentacles.
4. Food and oxygen enter the bodies of sponges as water passes through their pores. Cnidarians sting their prey with nematocysts and then pull the prey into their mouths. Worms that are parasites use the nutrients they obtain from a host.

Mollusks with radula file off bits of plant material and swallow the plant bits. Students will likely include other examples discussed in the chapter as well.
5. Moving ocean water carries oxygen into the sponge through its pores. Carbon dioxide is removed through the pores by a similar process.
6. Sponges are used for cleaning, are a source of antibiotics, and provide homes and food for other organisms.

Chapter Review

Content Review

Multiple Choice

Choose the letter of the answer that best completes each statement.

1. The largest and most general group in the classification system is the
   a. phylum.
   b. kingdom.
   c. species.
   d. class.
2. Organisms that can make their own food are called
   a. protists.
   b. heterotrophs.
   c. fungi.
   d. autotrophs.
3. Which is an invertebrate?
   a. lion
   b. human
   c. dog
   d. jellyfish
4. Which animal is a member of the phylum Porifera?
   a. sponge
   b. coral
   c. planarian
   d. squid
5. Cnidarians have
   a. pores.
   b. a mouth and an anus.
   c. mantles.
   d. nematocysts.
6. Which animal is a flatworm?
   a. hydra
   b. sea anemone
   c. planarian
   d. jellyfish
7. The animal that causes trichinosis is a
   a. roundworm.
   b. flatworm.
   c. segmented worm.
   d. cnidarian.
8. Tapeworms are
   a. sponges.
   b. flatworms.
   c. cnidarians.
   d. mollusks.
9. All mollusks have
   a. outer shells.
   b. soft bodies.
   c. nematocysts.
   d. spicules.
10. Clams are
    a. cnidarians.
    b. mollusks.
    c. flatworms.
    d. roundworms.

True or False

If the statement is true, write "true." If it is false, change the underlined word or words to make the statement true.

1. Animals are multicellular <u>autotrophs</u>.
2. <u>Invertebrates</u> have no backbone.
3. All sponges have <u>pores</u>.
4. Nematocysts are found in <u>mollusks</u>.
5. Flatworms are <u>nematodes</u>.
6. In <u>segmented worms</u>, the mantle produces the shell.
7. An oyster is an example of a <u>tentacled</u> <u>mollusk</u>.
8. The octopus is a <u>cnidarian</u>.

Concept Mapping

Complete the following concept map for Section 1–1. Refer to pages C6–C7 to construct a concept map for the entire chapter.

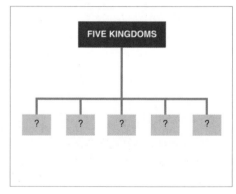

Concept Mastery

Discuss each of the following in a brief paragraph.

1. Why do you think it is important to study animals?
2. List the five kingdoms of living things and give a description of each.
3. Describe the similarities and differences among the three groups of mollusks.
4. List the different methods sponges, cnidarians, worms, and mollusks use to get food.
5. Describe how sponges take in oxygen and give off carbon dioxide.
6. In what ways are sponges useful?
7. Compare a medusa and a polyp.
8. Why are tapeworms parasites?
9. How are mollusks useful?
10. Compare sexual reproduction and asexual reproduction.

Critical Thinking and Problem Solving

Use the skills you have developed in this chapter to answer each of the following.

1. **Making charts** Construct a chart in which you list each phylum of invertebrates discussed in this chapter, the major characteristics of the phylum, and two examples from each phylum.
2. **Applying concepts** Why is it that scientists do not classify animals by what they eat or where they live?
3. **Making generalizations** In what ways are earthworms beneficial to humans?
4. **Designing an experiment** The eyes of a squid are similar in structure to the eyes of a vertebrate. Design an experiment to determine whether or not squids are able to see color. Include a hypothesis, control, and variable.
5. **Relating cause and effect** People with tapeworms eat a lot but still feel hungry and tired. Why?
6. **Making predictions** Would it be safe to eat clams from polluted water? Give a logical reason for your answer.
7. **Relating facts** Explain this statement: In a classification system, an organism is identified by showing what it is not.
8. **Using the writing process** Imagine that you could shrink down to a size small enough to fit inside an earthworm. Describe the adventures you and the earthworm would have in a garden one summer day.

7. A polyp is vase-shaped; a medusa is bowl-shaped. Both are body forms for cnidarians.
8. They live in the bodies of other organisms and take their food from their hosts.
9. Mollusks are a source of food for many organisms, including humans.
10. In sexual reproduction, a new organism is formed from the joining of a male and female cell. In asexual reproduction, a single organism produces the offspring. One type of asexual reproduction is budding.

CRITICAL THINKING AND PROBLEM SOLVING

1. Charts should be consistent with the information presented in the chapter.
2. Such a system would not help delineate all living things from one another because many living things live in similar areas and eat similar food substances. For example, both cows and horses eat grass and may live in pastures, but they are different organisms.
3. Earthworms keep the soil healthy by creating passages for air and depositing

fertilizer. Healthy soil will produce better gardens.
4. Investigations will vary but will likely include the idea of using a colored food source to attract the squid as opposed to a food source that is not colored.
5. The tapeworm uses much of the nutrients from the food people eat, and people do not actually get the nutrients they need, despite overeating.
6. No, because clams pass water through their bodies and some of the pollutants will build up in the clams.
7. Accept all logical responses. In order for a classification system to be useful, it must be clear whether an individual organism does or does not belong to a given category.
8. Accept all logical and well-written responses. The activities of the earthworm should be consistent with information provided in this chapter.

KEEPING A PORTFOLIO

You might want to assign some of the Concept Mastery and Critical Thinking and Problem Solving questions as homework and have students include their responses to unassigned questions in their portfolio. Students should be encouraged to include both the question and the answer in their portfolio.

ISSUES IN SCIENCE

The following issues can be used as springboards for discussion or given as writing assignments:
1. Why do you think scientists increased the number of categories of classifying from kingdom and species to include phylum, class, order, family, and genus?
2. Recently, scientists have discovered that certain kinds of bacteria are quite different in their chemical makeup from other monerans. They have proposed that a sixth kingdom called Archaebacteria be created for these bacteria because they are so different. Other scientists oppose this idea. Some argue that even five kingdoms are too many. If we keep inventing new kingdoms, classification systems may become too complex. What is your opinion? Should a sixth kingdom be established?

Chapter 2   ARTHROPODS AND ECHINODERMS

SECTION	HANDS-ON ACTIVITIES
2–1 Arthropods: The "Joint-Footed" Animals pages C40–C47 Multicultural Opportunity 2–1, p. C40 ESL Strategy 2–1, p. C40	**Student Edition** ACTIVITY (Discovering): The Flour Beetle, p. C44 ACTIVITY (Discovering): The Life of a Mealworm, p. C47 LABORATORY INVESTIGATION: Investigating Isopod Environments, p. C58 ACTIVITY BANK: Off and Running, p. C156 ACTIVITY BANK: Spinning Webs, p. C157 **Teacher Edition** Observing Crayfish, p. C38d
2–2 Insects: The Most Numerous Arthropods pages C48–C54 Multicultural Opportunity 2–2, p. C48 ESL Strategy 2–2, p. C48	**Student Edition** ACTIVITY BANK: How Many Are Too Many? p. C159 **Teacher Edition** Observing Eye Structure, p. C38d **Activity Book** ACTIVITY BANK: Collecting Soil Critters, p. C155 **Laboratory Manual** Observing Organisms in Soil, p. C41 Investigating an Ant Society, p. C51 The Grasshopper, p. C55
2–3 Echinoderms: The "Spiny-Skinned" Animals pages C54–C57 Multicultural Opportunity 2–3, p. C54 ESL Strategy 2–3, p. C54	**Laboratory Manual** The Starfish, p. C63
Chapter Review pages C58–C61	

OUTSIDE TEACHER RESOURCES

Books

Arnett, R. *Simon & Schuster's Guide to Insects*, Simon & Schuster.

Blaney, W. M. *How Insects Live*, E. P. Dutton.

Causey, Don. *Killer Insects*, Watts.

Johnson, Sylvia A. *Wasps*, Lerner.

Horton, C. *Insects*, Gloucester.

Lerner, Carol. *Pitcher Plants: The Elegant Insect Traps*, Morrow.

Patent, Dorothy Hinshaw. *Butterflies and Moths, How They Function*, Holiday.

Patent, Dorothy Hinshaw, and Paul C. Schroeder. *Beetles and How They Live*, Holiday.

Whalley, Paul. *Butterfly and Moth*, Alfred Knopf.

Audiovisuals

Ants and How They Live, video, AIMS Media Film and Video

Crustaceans, film, Encyclopaedia Britannica Education

Insects, film, Encyclopaedia Britannica Education

OTHER ACTIVITIES	MEDIA AND TECHNOLOGY
Student Edition Activity (Reading): A Silky Story, p. C42 **Activity Book** Activity: Studying Spider Webs, p. C37 **Review and Reinforcement Guide** Section 2–1, p. C27	**English/Spanish Audiotapes** Section 2–1
Student Edition Activity (Writing): What Kind of Insect Is It? p. C52 **Activity Book** Chapter Discovery: Observing Insect Behavior and Communication, p. C25 Activity: Insect or Arachnid? p. C29 Activity: Social Insects, p. C31 **Review and Reinforcement Guide** Section 2–2, p. C33	**Interactive Videodisc** Insects: Little Giants of the Earth On Dry Land: The Desert Biome **English/Spanish Audiotapes** Section 2–2
Activity Book Activity: Examining the Role of Starfishes in an Aquatic Community, p. C33 **Review and Reinforcement Guide** Section 2–3, p. C35	**Video** Echinoderms, Mollusks, and Arthropods (Supplemental) **English/Spanish Audiotapes** Section 2–3
Test Book Chapter Test, p. C29 Performance-Based Tests, p. C109	**Test Book** Computer Test Bank Test, p. C35

*All materials in the Chapter Planning Guide Grid are available as part of the Prentice Hall Science Learning System.

Life Cycle of the Honeybee, video, National Geographic
Spiders and How They Live, video, AIMS Film and Video
The World of Insects, video, National Geographic

Chapter 2 ARTHROPODS AND ECHINODERMS

CHAPTER OVERVIEW

In terms of sheer numbers, variety of species, and diversity of habitats, arthropods, a phylum of invertebrates, are the most successful animals on Earth. From their first appearance more than 600 million years ago, arthropods have been adapting and evolving into the organisms we know today. Arthropods include crustaceans, such as lobsters and isopods; millipedes and centipedes; arachnids, such as spiders and ticks and mites; and insects. Arthropods can both benefit and harm human and other species. They may be predators that eliminate harmful organisms from the environment, parasites that can cause disease, decomposers, or symbionts. Despite their roles—harmful or beneficial—arthropods are encountered by all other animals at some point in their lives.

The common features shared by all arthropods are exoskeletons, segmented bodies, and jointed legs. Arthropods develop through complete or incomplete metamorphosis, and they molt, or shed, their exoskeletons as they grow. Insects, the largest group of arthropods, live as individuals or in colonies in which all members contribute to the good of the community.

Echinoderms are another phylum of invertebrates. As a group, echinoderms have many unusual features. They appear to have a crusty and spiny exoskeleton. They are also characterized by their unique water vascular systems and tube feet. This unique system enables echinoderms to build up and release pressure for movement and gripping things. Some echinoderms have the ability to regenerate, or replace parts. Members of the phylum include the starfish, the sand dollar, and the sea cucumber. Echinoderms are sea creatures, and a number of colorful and varied organisms make up the phylum.

2–1 ARTHROPODS: THE "JOINT-FOOTED" ANIMALS
THEMATIC FOCUS

The purpose of this section is to introduce students to the invertebrates known as arthropods. Students first identify the characteristics—exoskeletons, segmented bodies, and jointed legs—common to all arthropods. They then examine specific members of the phylum including the crustaceans, millipedes and centipedes, and arachnids. In their examinations, they identify similarities and differences in members of the phylum.

The themes that can be focused on in this section are energy, evolution, scale and structure, and unity and diversity.

Energy: Arthropods, like other animals, obtain energy by eating plants or animals. Through the digestive process, these food sources are converted into energy-producing substances.

*Evolution: The success of the arthropods is due in a large part to evolutionary changes over about 600 million years. The changes have enabled arthropods to adapt to life in the sea, in the air, and on the land.

*Scale and structure: Because of their exoskeletons, most arthropods are incapable of growing very large. Arthropods that live in the sea tend to be larger because buoyancy helps to support the weight of the animal.

*Unity and diversity: Although all arthropods have segmented bodies and jointed legs, the number of segments and of jointed appendages varies from group to group. These differences in number of body segments and jointed legs can help people distinguish one group of arthropods from another.

PERFORMANCE OBJECTIVES 2–1

1. **Name major characteristics of arthropods.**
2. **State advantages and disadvantages of exoskeletons.**
3. **Identify characteristics of crustaceans, millipedes and centipedes, and arachnids and cite examples of each.**

SCIENCE TERMS 2–1

exoskeleton p. C41
molting p. C41

2–2 INSECTS: THE MOST NUMEROUS ARTHROPODS
THEMATIC FOCUS

The purpose of this section is to introduce students to the largest group of arthropods, the insects. Students first identify physical characteristics common to all insects. They then examine the stages of development of insects through complete and incomplete metamorphosis. Finally, students investigate the behavioral patterns of insects including the mating practices, social organizations, and defensive techniques.

The themes that can be focused on in this section are patterns of change, systems and interactions, unity and diversity, and stability.

*Patterns of change: As insects develop, they go through a series of changes through complete and incomplete metamorphosis. In complete metamorphosis, the insect's appearance changes from one stage to another. In incomplete metamorphosis, the insect changes by maturing sexually and acquiring wings as it reaches adulthood.

Systems and interactions: As arthropods, insects respond to and interact with their environment in ways that help them gather food, reproduce, and protect themselves. Some insects, such as bees and ants, belong to societies in which members have specific functions to benefit the entire colony. Other insects live out their lives as individuals that do not compete with other members of the group for food but meet to reproduce.

*Unity and diversity: Insects perform the basic life functions in different ways. Some live individually, whereas others live in organized colonies. Methods of reproduction, food sources, and internal structures vary from one group to another, but all groups share common structural characteristics.

Stability: The organization of social insects has remained stable. Each member of a community has a role to play within the community. The life functions of insects enable them to maintain a stable internal and external environment.

PERFORMANCE OBJECTIVES 2–2

PERFORMANCE OBJECTIVES 2–2

1. **Describe the distinguishing characteristics of insects.**
2. **Identify the stages of an insect's metamorphosis.**
3. **Detail the social structure of a bee colony.**
4. **Cite examples of insect behaviors including defense mechanisms.**

SCIENCE TERMS 2–2

metamorphosis p. C49
larva p. C49
pupa p. C50
pheromone p. C51

2–3 ECHINODERMS: THE "SPINY-SKINNED" ANIMALS

THEMATIC FOCUS

The purpose of this section is to introduce students to invertebrates known as echinoderms. Students first identify characteristics common to all echinoderms. They then study the starfish in detail as an example of echinoderms. They also learn about other members of this phylum of invertebrates.

The themes that can be focused on in this section are scale and structure, systems and interaction, and patterns of change.

*Scale and structure: All echinoderms have spiny skins, an internal skeleton, a five-part body, a water vascular system, and tube feet. The tube feet enable the echinoderms to move, hold on to objects, and feed. The water vascular system expands and contracts to enable the feet to operate.

Systems and interactions: Because of the water vascular systems of echinoderms, all species live in the water. These systems are vital for obtaining oxygen and food.

*Patterns of change: Many echinoderms are able to regenerate, or develop new individuals or organs that have been harmed. This ability helps to protect the members of the phylum and ensures their continued existence.

PERFORMANCE OBJECTIVES 2–3

1. **Identify the distinguishing features of echinoderms.**
2. **Describe the functions of a starfish's tube feet.**
3. **Give examples of echinoderms that differ in appearance from starfishes.**

SCIENCE TERMS 2–3

water vascular system p. C54
tube foot p. C54

Discovery Learning

TEACHER DEMONSTRATIONS MODELING

Observing Crayfish

Obtain a few live crayfishes for students to observe. Some students may be able to collect these from local streams, ponds, or marshes. Crayfish are also sometimes sold at bait stores as "soft craws," and they can also be purchased from biological supply companies.

Place a few specimens in beakers or battery jars at several locations around the room for small groups of students to examine. Put only a small amount of water in the containers. Ask that groups make some specific observations.

• **How many antennae (feelers) does the crayfish have on its head?** (Students should notice that it has two long antennae that bend backward over the body. Mention that these antennae are the crayfish's organ of touch, taste, and smell. Students should also notice the pair of short, branched antennules. These antennules serve as the organs of balance and hearing.)

Provide some small pieces of meat. Ask the groups to place these near the crayfish's head. If the crayfish feeds, ask them to note how it obtains and eats the meat.

• **Does the crayfish chew its food or does it swallow it whole?** (The group should notice that the crayfish uses its appendages near the mouth to hold the food as it chews the food with its mandibles, or jaws.)

Tell the groups to gently place the crayfish on the table top and watch it walk. Caution students to keep their fingers away from the claws of the crayfish.

• **How many legs does the crayfish use for walking?** (Unless legs are missing, the crayfish uses its four pairs of walking legs.)

Next have students place the crayfish in an aquarium or a plastic dishpan and observe it swim.

• **Describe how the crayfish moves in the water.** (Students should note that the small appendages, called swimmerets, and the flipperlike tail sections are used to move in water. The swimmerets are located on the lower side of the abdomen.)

Summarize by discussing the different kinds of specialized appendages that a crayfish has and identifying the crayfish as a member of the Arthropoda phylum.

Observing Eye Structure

Obtain a preserved grasshopper (a high-school biology teacher may be able to provide one) and set up a demonstration for students to observe the makeup of a compound eye. To do this, use a scalpel or razor blade to cut a thin section from the eye. Then lay the piece flat on the microscope slide and put it under a stereomicroscope for students to observe. After everyone has had the opportunity to view the eye, ask the following questions:

• **What does an insect's eye look like?** (Students should indicate that it is made up of many separate sections.)

• **What is this type of eye called?** (A compound eye.)

Point out that each section is a separate eye with its own lens and nerve to the brain. Some insects' eyes are made up of almost 30,000 lenses. With such eyes, insects view their world as a mosaic pattern of dots. Nevertheless, they are able to see to the sides and rear as well as straight ahead.

• **What advantage is it for an insect to be able to see in all directions?** (They are able to see enemies approaching from any direction.)

CHAPTER 2
Arthropods and Echinoderms

INTEGRATING SCIENCE

This life science chapter provides you with numerous opportunities to integrate other areas of science, as well as other disciplines, into your curriculum. Blue numbered annotations on the student page and integration notes on the teacher wraparound pages alert you to areas of possible integration.

In this chapter you can integrate life science and evolution (p. 40), social studies (p. 41), language arts (pp. 41, 42, 54), mythology (p. 44), life science and health (p. 47), physical science and light (p. 49), physical science and aerodynamics (p. 57).

SCIENCE, TECHNOLOGY, AND SOCIETY/COOPERATIVE LEARNING

A tiny arachnid is spreading disease and increasing alarm across the country. Lyme disease is caused by a spirochete spread from one organism to another by a poppy-seed-sized tick that is infected with the bacterium. Lyme disease is one of several diseases spread by ticks—but because it is hard to diagnose and its symptoms are so debilitating, Lyme disease has caused much fear and anxiety in people who live in tick-infested areas.

The first problem with diagnosing Lyme disease is that many people do not realize that they have been bitten! The tiny tick may crawl around on a person for hours, bite the individual, and be gone before the person ever realizes that he or she has been bitten. Once infected, a person may develop a variety of symptoms: a characteristic bull's-eye rash, flu-like symptoms, pain and swelling in the joints, and, occasionally, cardiac trouble, facial paralysis, and numbness of parts of the body. Once diagnosed, the person is treated with antibiotics such as tetracycline and penicillin. Cases diagnosed after a longer period of time often require several treatments with very expensive, high-powered antibiotics, and even

INTRODUCING CHAPTER 2

DISCOVERY LEARNING

▶ *Activity Book*
Begin teaching the chapter by using the Chapter 2 Discovery Activity from the *Activity Book*. Using this activity, students will discover some methods used by insects to communicate with one another.

USING THE TEXTBOOK

After students examine the picture on page C38 and read the caption on page C39, ask if they have ever seen that many fireflies in one place. Explain that some people call fireflies "lightning bugs." Then have students read the chapter-opening text.
• **Why is a firefly's light considered a marvel of nature?** (It is light without much heat.)
• **How are fireflies like other insects?**

Arthropods and Echinoderms

Guide for Reading

After you read the following sections, you will be able to

2–1 Arthropods: The "Joint-Footed" Animals
- Describe the characteristics of arthropods.
- Identify the major groups of arthropods.

2–2 Insects: The Most Numerous Arthropods
- Describe the characteristics of insects.
- Explain how insects behave.

2–3 Echinoderms: The "Spiny-Skinned" Animals
- Describe the characteristics of echinoderms.

As darkness falls on a warm summer evening, fireflies light their "lanterns." The temperature has to be warm enough and the late-day light dim enough for them to be seen. If these conditions prevail, the tiny magicians light up the night worldwide.

A firefly's light is one of nature's marvels. It is light with almost no heat—a feat humans have yet to achieve. In fact, a firefly's light is cooler than the warm night air that surrounds it.

The greatest firefly show in the world occurs during summer evenings in Thailand. Male fireflies bunch together on certain trees that line the rivers. Flashing 120 times a minute, the male fireflies soon regulate their flashes so that at one instant there is total blackness, and at the next instant total illumination—over and over again!

Fireflies (which are actually beetles, not flies) belong to a phylum of invertebrates called arthropods. In this chapter you will read about this phylum of fascinating creatures. And you will also be introduced to another phylum of invertebrates, the echinoderms, which includes the lovely yet somewhat dangerous purple sea urchin.

Journal *Activity*

You and Your World Think of a place near your home or school where animals may live. It can be a schoolyard, a backyard, an empty lot, a park, or even an alley. It should have some soil, rocks, green plants, and a source of moisture. Write a detailed description of the area in your journal. Draw a picture of it. Then visit the area and see if your description was accurate. Make any necessary changes in your journal.

The Thailand night is illuminated by an awesome firefly show.

then some of the symptoms may be irreversible.

Trying to fight this insect-borne disease is proving to be very difficult. Environmental strategies for controlling the tick have been unsuccessful—widespread spraying of chemicals is ineffective and dangerous. Eliminating the tick's host, the white-tailed deer, is opposed by animal lovers. Experts feel that public education is the most effective weapon available for fighting Lyme and other tick-borne diseases.

To protect themselves from ticks, people should take the following precautions when in areas infested by ticks: Wear long-sleeved shirts and long pants; tuck shirts into pants and pants into socks; wear light-colored clothing to make the ticks easier to spot; check often for ticks; use insect repellant; check pets for ticks.

Cooperative learning: Using preassigned lab groups or randomly selected teams, have groups complete one of the following assignments.

• Design a series of signs to be used along hiking trails in national parks to remind hikers of ways to protect themselves from ticks. The signs should use visual reminders as opposed to a long written message that the hikers might not stop and read.

• Because public education is the best weapon against Lyme disease, prepare an informative pamphlet that alerts people to the dangers of tick-borne diseases, identifies methods for protecting oneself from ticks, and lists symptoms of Lyme and other tick-borne diseases.

See Cooperative Learning in the *Teacher's Desk Reference.*

JOURNAL ACTIVITY

You may want to use the Journal Activity as the basis of a class discussion. Before students examine the place they have chosen, encourage them to look closely for all types of animals including insects, spiders, and other small creatures. After they visit the place, discuss whether they saw creatures they did not expect to see or whether they saw fewer animals than they expected. Have students place their Journal Activity in their portfolio.

(Students might suggest that they have wings, six legs; they are invertebrates called arthropods.)

• **What is the purpose of a firefly's light?** (A male shines its light to attract a female partner.)

• **Do you think the male firefly is able to control its light? Why or why not?** (Because the fireflies in Thailand regulate their lights to shine at the same time, students may state that the fireflies are able to control their lights.)

Explain that generating a light to attract a female is a behavioral characteristic of fireflies. This behavior, however, is dependent on the physical features of the firefly.

• **Is the behavior of generating a light to attract a partner a characteristic common to all insects?** (No, other insects do not exhibit this behavior.)

• **How might other insects attract partners?** (Accept all logical answers.)

2-1 Arthropods: The "Joint-Footed" Animals

Guide for Reading

Focus on these questions as you read.

▶ What are the main characteristics of arthropods?

▶ What are the characteristics of each of the four groups of arthropods?

Figure 2–1 *The spider crab (top left), millipede (top right), daddy longlegs (bottom right), and weevil and ant (bottom left) are members of the largest and most diverse group of invertebrates: the arthropods. What three characteristics are shared by all arthropods?* ❶

2–1 Arthropods: The "Joint-Footed" Animals

The phylum of invertebrates that contains the greatest number of species is the phylum Arthropoda (ahr-THRAHP-uh-duh). Members of this phylum are called arthropods. To date, more than 1 million species of arthropods have been described. Scientists estimate, however, that the total number of arthropod species may be as high as 1 billion billion, or 1,000,000,000,000,000,000! Arthropods live in air, on land, and in water. Wherever you happen to live, you can be sure arthropods live there too. Arthropods are our main competitors for food. In fact, if left alone and unchecked, they could eventually take over the Earth!

Why are there so many arthropods? One reason is that they have been evolving (changing) on Earth for more than 300 million years. During this time, they have developed certain characteristics that allowed them to become so successful. Of these characteristics, three are common to all arthropods. **The three characteristics shared by all arthropods are an exoskeleton, a segmented body, and jointed appendages.**

legs—have developed over time. The exoskeleton of the arthropod is a rigid outer shell made of chitin.

• **How is an exoskeleton beneficial to an arthropod?** (The exoskeleton serves as protection against predators. It helps to prevent water loss from the body. Thus, some arthropods are capable of living in dry environments. The exoskeleton also serves as a point for muscle attachment, making efficient movement of the joints possible.

The most striking characteristic of arthropods is the **exoskeleton.** An exoskeleton is a rigid outer covering. The exoskeletons of many land-dwelling arthropods are waterproof. Such exoskeletons limit the loss of water from the bodies of arthropods, thus making it possible for them to live in remarkably dry environments such as deserts. In some ways, an exoskeleton is like the armor worn by knights in the Middle Ages as protection in battle. One drawback of an exoskeleton, however, is that it does not grow as the animal grows. So the arthropod's protective suit must be shed and replaced from time to time. This process is called **molting.** While the exoskeleton is replacing itself, the arthropod is more vulnerable to attack from other animals.

Like annelids (segmented worms), arthropods have segmented bodies. This characteristic strongly suggests that annelids and arthropods evolved from a common ancestor. The body of most arthropods, however, is shorter and has fewer segments than an annelid's.

Although the phylum name, Arthropoda, comes from the Greek words meaning jointed legs, it is not just legs that enable arthropods to move. The appendages characteristic of arthropods include antennae, claws, walking legs, and wings as well. See Figure 2–4 on page 42.

An arthropod has an open circulatory system, or one in which the blood is not contained within small tubes. Instead, the blood is pumped by a heart throughout the spaces within the arthropod's body.

Figure 2–2 *In order to increase in size, arthropods must molt, or shed their exoskeletons. This adult cicada is emerging from its exoskeleton, which it has outgrown.*

Figure 2–3 *Some arthropods, such as the Hawaiian lobster and dust mite have hard, tough exoskeletons. Other arthropods, such as the Promethea moth caterpillar, have flexible exoskeletons. What is an exoskeleton?* ②

C ■ 41

ANNOTATION KEY

Answers

① An exoskeleton, a segmented body, and jointed appendages. (Relating facts)

② A rigid outer covering. (Applying definitions)

Integration

① Life Science: Evolution. See *Evolution: Change Over Time*, Chapter 1.

② Social Studies

③ Language Arts

BACKGROUND INFORMATION

ARTHROPOD EVOLUTION

Until recently, most biologists believed that arthropods were evolved from a common ancestor because arthropods share such easily observable characteristics as chitinous exoskeletons, jointed legs, and segmented bodies. These similar characteristics, however, may be the result of convergent evolution, or adaptive evolution of physical characteristics of species living in the same environment, rather than the result of a shared ancestry. Embryological research has shown that there are at least two and probably four distinctly separate evolutionary lines within the phylum.

• **What is a disadvantage of an exoskeleton?** (An exoskeleton limits an arthropod's growth. To grow, the arthropod must shed, or molt, the exoskeleton. While molting, an arthropod is more vulnerable to predators.)

● ● ● ● **Integration** ● ● ● ●

Use the information about the development of the arthropods to integrate evolutionary concepts into your science lesson.

Use the comparison between exoskeletons and knight's armor to integrate social studies concepts into your science lesson.

Use the discussion of the etymology of *Arthropoda* to integrate language arts concepts into your science lesson.

GUIDED PRACTICE

▶ *Laboratory Manual*

Skills Development

Skills: Making comparisons, identifying characteristics, making observations, classifying, drawing conclusions

At this point assign the Laboratory Investigation called Observing Organisms in the Soil in the *Laboratory Manual.* Through this investigation students will observe a number of invertebrates from surface-soil and deep-soil sources.

ACTIVITY READING

A SILKY STORY

Skill: Reading comprehension

Charlotte's Web by E. B. White is an enchanting story about a spider who devotes herself to praising Wilbur the Pig to save him from slaughter. As students read the story, ask them to consider why people did not recognize Charlotte as the remarkable individual in the story. Have them consider how this attitude about Charlotte reflects people's attitudes about most arthropods.

Integration: Use this Activity to integrate language arts into your science lesson.

FACTS AND FIGURES

CRUSTACEANS

In general, crustaceans are characterized by their hard exoskeletons, two pairs of antennae, and mouthparts called mandibles. The antennae of most crustaceans are used primarily as feelers, or sense organs; in others, they also are used in filter feeding; in still others, they are used to help move the organism. Crustaceans, especially those that live in the sea, are often larger than other arthropods.

BACKGROUND INFORMATION

CRUSTACEAN REPRODUCTION

In most crustaceans, mating occurs when males use their first swimming legs, or flipperlike appendages, to transfer sperm to the seminal receptacles of females. Lobsters, which greatly resemble large marine crayfish, mate just after the female molts as the ocean warms up in the spring. After the eggs are fertilized, the female attaches hundreds of them to the swimming legs under her abdomen with a sticky secretion. There she carries them for almost nine months as they develop through winter and early spring of the next year, at which time they hatch into free-swimming larvae.

Figure 2–4 *Arthropod appendages include the antennae of a harlequin beetle, the claws of a fiddler crab, the legs of a praying mantis, and the wings of a green lacewing fly.*

Activity Bank

Off and Running, p.156

A Silky Story

1 You may want to read a wonderful story about an exceptional spider named Charlotte who weaves messages into her web. The book is entitled *Charlotte's Web,* and it was written by E. B. White.

42 ■ C

Although an arthropod's blood carries food throughout its body, it does not carry oxygen. Oxygen is carried to all the cells by one (sometimes two) of three basic respiratory (breathing) organs—gills, book lungs, and a system of air tubes.

Arthropods reproduce sexually. That means there are two parents, a male and a female. The male produces sperm, and the female produces eggs. In most arthropods, the sperm and the egg unite inside the body of the female.

Various animals make up this phylum. They include crustaceans (kruhs-TAY-shuhnz), centipedes and millipedes, spiders and their relatives, and insects.

Crustaceans

Do you see the two eyes peering out at you from the shell in Figure 2–5? These eyes belong to an animal known as a hermit crab. A hermit crab is a crustacean that lives in shells discarded by other water-dwelling animals such as mollusks. A crustacean is an arthropod that has a hard exoskeleton, two pairs of antennae, and mouthparts that are used for crushing and grinding food. Crustaceans include crabs, lobsters, barnacles, and shrimp.

2-1 (continued)

CONTENT DEVELOPMENT

Unlike most arthropods, the exoskeletons of some large crustaceans contain calcium carbonate. This substance makes the shells of crustaceans, such as lobsters and crabs, hard and stony. Point out that calcium carbonate is the chemical compound from which mollusks form their shells and from which limestone rocks on land are made. Another characteristic of crustaceans is that they have two pair of antennae. The mouthparts, or mandibles, of the crustaceans vary from species to species. In many species, including crayfish, the mandibles are short, heavy structures used to bite and grind foods. In some species, the mandibles are bristly structures used for filter feeding. In filter feeding, food such as tiny plants is strained from the water taken in by the crustaceans. In still other species, mandibles are probelike structures that find

The body of a crustacean is divided into segments. A pair of appendages is attached to each segment. The type of appendage varies, depending on the crustacean. Crabs, for example, have claws. The claws of some crabs are so strong that they can be used to open a crab's favorite food, a coconut. Crabs also have walking legs and antennae, which are some other examples of appendages.

Crustaceans such as crabs are able to regenerate (regrow) certain parts of their body. The stone crab, which lives in the waters off the coast of Florida, can grow new claws. This is an important characteristic for a stone crab because its claws are considered particularly tasty by people. When a stone crab is caught, one of its claws is broken off and the stone crab is returned to the water. In about a year's time, the missing claw is regenerated. If a crab is caught again, that claw may once again be removed.

Most crustaceans live in watery environments and obtain oxygen from the water through special respiratory organs called gills. Even the few land-dwelling crustaceans have gills. Such crustaceans, however, must live in damp areas in order to get oxygen.

Figure 2–5 *Water-dwelling crustaceans include the hermit crab and goose neck barnacles.*

Figure 2–6 *The diagram shows some of the internal and external structures of a crayfish.*

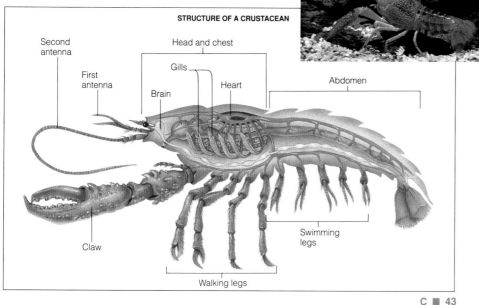

STRUCTURE OF A CRUSTACEAN

Second antenna

First antenna

Head and chest

Gills

Brain

Heart

Abdomen

Claw

Swimming legs

Walking legs

C ■ 43

INTEGRATION
HOME ECONOMICS

Crustaceans, such as crabs, lobsters, shrimp, and crayfish, are widely used as a source of food for humans. Fine restaurants feature them in numerous specialties, and they are common staples in many households. Ask students to list as many dishes as they can think of in which crustaceans are used. From the individual lists, compile a class list of crustacean foods on the chalkboard. If any students are interested in cooking, they may want to share the recipes for some of their favorite crustacean dishes. You might also have students investigate the health risk crustaceans may present for some people. For example, people with high blood pressure should avoid many crustaceans because of their high salt content.

FACTS AND FIGURES
HORSESHOE CRABS

Despite their name, horseshoe crabs are not true crabs, and they are therefore not classified as crustaceans. They are actually the only living representatives of a group of arthropods whose other members are now extinct. The closest relatives of horseshoe crabs appear to be scorpions and spiders rather than crustaceans.

and pick up foods or needlelike structures that suck fluids from a food source.

GUIDED PRACTICE

Skills Development

Skills: Making observations, predicting, hypothesizing, inferring, drawing conclusions

At this point you may want students to complete the in-text Laboratory Investigation. In the investigation students hypothesize about the environmental preferences of the isopod, a crustacean that lives on land.

ENRICHMENT

Crustaceans differ from most other arthropods in the number and degree of specialization of their appendages. Encourage interested students to conduct research on a crayfish's appendages and their specialized functions. Have them prepare a large drawing with the appendages and their functions.

BACKGROUND INFORMATION

CHILD CARE IN SPIDERS

Female spiders usually lay eggs in a small cocoon spun from silk. In some, such as members of the genus *Theridion,* the young live on the mother's web for a month or so after hatching. When the mother captures prey, she signals to the young by strumming the web with her legs. When danger threatens, she rubs the web a different way, and the young scurry to the protection of their cocoon. Once they are a few weeks old, most spiders live alone. Some baby spiders leave the nest by climbing onto a tall plant and releasing a long silk thread. When a strong wind picks up the thread, the spider lets go of its perch and sails off in the wind. This behavior, called ballooning, can carry the baby spider for hundreds of kilometers to a new, possibly less crowded, territory.

Figure 2–7 *Although a centipede (top) and a millipede (bottom) look very much alike, they do differ in some ways. What are two ways in which centipedes and millipedes differ?* ●

Centipedes and Millipedes

Centipedes and millipedes are arthropods that have many legs. What distinguishes one from the other is that centipedes have one pair of legs in each segment, whereas millipedes have two pairs of legs in each segment.

If you were an earthworm crawling through the soil, you would certainly be aware of another difference between centipedes and millipedes. Millipedes live on plants and thus would simply pass you by. Millipedes are shy creatures. When disturbed, millipedes may roll up into a ball to protect themselves. Centipedes, on the other hand, are carnivores, or flesh eaters. They are active hunters. To capture you, a centipede would inject poison into your body through its claws. (Another difference between centipedes and millipedes is that centipedes have claws and millipedes do not.)

Unlike many arthropods, the exoskeletons of centipedes and millipedes are not waterproof. To avoid excessive water loss, centipedes and millipedes are usually found in damp places such as under rocks or in soil.

Spiders and Their Relatives

There is a legend in Greek mythology about a young woman named Arachne (uh-RAK-nee) who ● challenged the goddess Athena to a weaving contest. When Arachne won the contest, Athena tore up Arachne's tapestry. Arachne hanged herself in sorrow. Athena then changed Arachne into a spider and Arachne's tapestry into a spider's web. Today spiders, as well as scorpions, ticks, and mites, are included in a group of arthropods called arachnids (uh-RAK-nihdz). As you can see, the word arachnid comes from the name Arachne.

The body of an arachnid is divided into two parts: a head and chest part and an abdomen part. Although arachnids vary in size and shape, they all have four pairs (8) of walking legs. So if you ever find a small animal you think is a spider, count its legs to make sure.

Spiders usually feed on insects. A few types of tropical spiders, however, can catch and eat small

---

2–1 (continued)

GUIDED PRACTICE

Skills Development

Skills: Making comparisons, defining terms

Mention that many people have difficulty distinguishing between centipedes and millipedes.

• **What does the prefix** *centi-* **mean?** (*Centi-* means 100.)

• **Does a centipede actually have 100 legs?** (The actual number of legs is usually fewer than 30.)

• **How many legs are on each segment of a centipede's body?** (Two legs.)

• **What does the prefix** *milli-* **mean?** (Students generally know that *milli-* means 1000.)

• **Do millipedes have 1000 legs?** (No, millipedes have two pairs of walking legs, or four walking legs per segment.)

• **Besides the number of legs, there is**

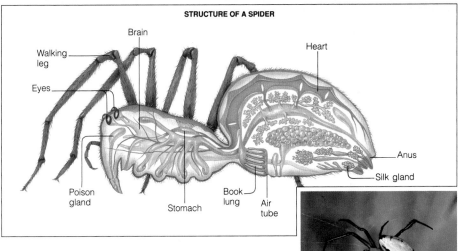

STRUCTURE OF A SPIDER

Brain
Walking leg
Eyes
Heart
Anus
Silk gland
Poison gland
Stomach
Book lung
Air tube

vertebrates such as hummingbirds. Spiders catch their prey in different ways. Many make webs of fine, yet very strong, flexible material called silk. Silk is secreted by special structures located in a spider's abdomen. Many spiders weave a new web every day. At night, the spiders eat the strands of that day's web, recycling the material the following day when they produce a new web. Although many spiders do not spin webs, they all produce silk. The silk that makes up the web is five times stronger than steel!

Some spiders hide from their prey and then suddenly jump out, taking the unsuspecting victim by surprise. For example, the trapdoor spider lives in a hole in the ground covered by a door made of silk. The door itself is hidden by soil and bits of plants. When an unsuspecting insect passes close to the trapdoor, the trapdoor spider jumps out and catches the unlucky victim.

Once a spider catches its prey, it injects venom, or poison, into the prey through a pair of fangs. Sometimes the venom kills the prey immediately. Other times it only paralyzes the prey. In this way, the spider can preserve living creatures trapped in its web for a time when it needs more food.

Most spiders get their oxygen by means of respiratory organs called book lungs. This is an appropriate name for these organs because the several sheets

Figure 2–8 *Spiders such as the orchard spider (top) and flower crab spider (bottom) belong to the group of arthropods called arachnids. Compare the structures that you see in the photographs with those shown in the diagram of the spider. How many legs do arachnids have?* ②

C ■ 45

FACTS AND FIGURES
THE POISON OF SPIDERS

The poison of only a few species of spiders is harmful to humans. The most dangerous is the bite of the black widow. This poisonous species occurs worldwide in warm areas, including much of the United States. The venom of the black widow is about 15 times more potent than that of the rattlesnake, but because the spider's supply is so small, death from a bite is rare. The injected poison, however, can cause severe pain and muscle paralysis. Only the female black widow bites. She is a large glossy-black spider that can be easily identified by a red hourglass marking on the underside of her abdomen.

pendages, called chelicerae, are modified as claws or fangs. In spiders, ducts from a pair of poison glands lead to the chelicerae. Most arachnids can produce silk to spin webs.

another important difference between centipedes and millipedes. On what do these animals feed? (Centipedes are predators that feed on insects; millipedes are scavengers that feed on dead vegetation.)

CONTENT DEVELOPMENT

Many people look on arachnids with fear or distate. Because some of them—especially ticks and mites—are parasites, such distaste is justified. Yet spiders, the most numerous arachnids, feed mainly on insects, and they play an important part in keeping the numbers of insects in check. In this way spiders are beneficial to humans.

Point out that arachnids differ from other arthropods in several ways. Arachnids do not have antennae on their heads, and rather than large compound eyes possessed by many arthropods, arachnids have eight simple eyes called ocelli. In arachnids, the first set of ap-

● ● ● ● **Integration** ● ● ● ●
Use the legend of Arachne to integrate Greek mythology into your science lesson.

ENRICHMENT

▶ *Activity Book*
Students will be challenged by the chapter activity called Spider Webs. In this activity students identify different types of webs that spiders spin and tell how the spiders use the webs to trap prey.

The arachnids that affect humans most directly are ticks and mites. Ticks, which are generally larger than mites, often live as parasites on humans and their domestic animals. Though a tick bite can be dangerous itself, more often the danger of the bite is from the wide variety of diseases that can be transmitted by these arachnids. Among the more dangerous tick-spread diseases are encephalitis, Rocky Mountain spotted fever, Lyme disease, and tularemia. Mites are a cause of a number of skin irritations, such as chigger bites and scabies in humans and mange on dogs and cats. Many human allergies are also thought to be caused by mites borne by dust particles.

FACTS AND FIGURES

THE WORLD'S LARGEST CENTIPEDE

The largest centipede is the giant scolpender that lives in Central America. It may reach a length of up to 30 centimeters and a width of 2.5 centimeters. Though it feeds mostly on insects, it is also fond of lizards and mice. In the tropics, several species of millipedes grow up to 30 centimeters long with a body width as great as 2 centimeters.

Figure 2-9 *Some spiders, such as the orb weaver spider, build webs to catch their prey. What material makes up the web?* ❶

of tissues that make up the structure resemble pages in a book. As air passes over the book lungs, oxygen is removed. Some spiders, however, have respiratory organs that form a system of air tubes. These air tubes are connected to the outside of the spider's body through small openings in its exoskeleton.

Arachnids live in many environments. Most spiders live on land. One interesting exception is a spider that lives under water inside bubbles of air that it carries down from the surface. When the air bubbles are used up, the spider returns to the water's surface for a fresh supply.

Scorpions are generally found in dry desert areas. Scorpions are active primarily at night. During the day, they hide under logs, stones, or in holes in the ground. People who enjoy camping must be careful

Figure 2-10 *Spiders capture their prey in several ways. The trapdoor spider (left) lies in wait and then leaps out to grab unlucky insects. The wolf spider (bottom right) hides in burrows in the sand waiting for its unsuspecting prey. The tarantula (top right) is large enough to catch and eat small invertebrates.*

2-1 (continued)

GUIDED PRACTICE

Skills Development

Skills: Making observations, manipulative, making prints

Some students may have an interest in making web prints of various species of spiders. To make these prints, they will need to obtain a can of either white spray enamel or plastic spray from an art sup-

ply store. They will also need a sheet of dark construction paper for each print to be made. Students first must chase the spider from the web and then coat both sides of the web by spraying it lightly from an angle. They place the construction paper on the web lightly, and cut away any part of the web extending beyond the edge of the paper. The web print should then be allowed to dry. A set of prints can be used to make an unusual bulletin board display.

CONTENT DEVELOPMENT

Mites and ticks are basically the same thing. Ticks are large mites that are characterized by recurved teeth on their chelicerae. Many are parasites living off hosts including humans, animals, and agricultural plants. Most are very small, less than 1 millimeter. In fact, one type of microscopic mite lives on the eyelashes of all humans. Some ticks, however, can be as large as 3 centimeters.

when they put on their shoes or boots in the morning: A scorpion may have mistaken the footwear for a suitable place in which to escape from the heat of the day! When scorpions capture prey, they hold it with their large front claws and, at the same time, inject it with venom through the stingers in their tails.

Ticks and mites live on other organisms. They may live on a plant and stay in one place, or they may live on an animal and travel wherever the animal travels. Like certain flatworms and roundworms, ticks and mites live off the body fluids of plants and animals. Some live by sucking juices from the stems and leaves of plants. Other ticks and mites are very tiny and live on insects. Many ticks suck blood from larger animals. In the process, they may spread disease. For example, Rocky Mountain spotted fever and Lyme disease are spread to people through the bites of ticks.

2–1 Section Review

1. What are the three main characteristics of arthropods?
2. What is an exoskeleton?
3. List four groups of arthropods.
4. What are some ways in which spiders and their relatives catch prey?

Critical Thinking—*Applying Concepts*
5. Blue crabs usually have hard shells. During certain times of the year, however, some blue crabs have thin, papery shells and are called soft-shell crabs. In terms of the life processes of arthropods, explain why these blue crabs have soft shells.

Figure 2–11 *Notice the stinger at the end of the scorpion's tail. The wood tick and the red velvet mites, which are devouring a termite, are arachnids.*

Activity Bank

Spinning Webs, p.157

ACTIVITY DISCOVERING

The Life of a Mealworm

1. Fill a clean 1-liter jar about one third full of bran cereal.
2. Place four mealworms in the jar.
3. Add a few slices of raw potato to the jar.
4. Shred a newspaper and place it loosely in the jar.
5. Cover the top of the jar with a layer of cheesecloth. Use a rubber band to hold the cheesecloth in place.
6. Observe the jar at least once a week for 4 weeks. Record all changes that take place in the mealworms.

How long was the mealworms' life cycle?

■ Do mealworms undergo complete or incomplete metamorphosis? Explain.

C ■ 47

ACTIVITY DISCOVERING
THE LIFE OF A MEALWORM

Discovery Learning

Skills: Making observations, manipulative, recording, applying, relating

Materials: 1-liter jar, bran cereal, raw potato, newspaper, cheesecloth, rubber band, 4 mealworms

Mealworms are not worms; they are insects. They are the larval stage of the grain beetle. As students may surmise, they live in stored meal, grain, or cereal. The complete life cycle of a mealworm may take four to six months.

Mealworms undergo complete metamorphosis. Evidence of complete metamorphosis is that mealworms are the larval stage of the adult grain beetle and do not resemble the adult.

● ● ● ● **Integration** ● ● ● ●

Use the information about diseases spread to humans by mites and ticks to integrate concepts of health into your lesson.

REINFORCEMENT/RETEACHING

Direct students to make a summary table with these three headings at the top: Arthropod Group, Characteristics, Examples. Have them complete the chart with information about crustaceans, millipedes and centipedes, and arachnids.

INDEPENDENT PRACTICE

Section Review 2–1

1. Exoskeleton, segmented body, jointed legs.
2. A rigid outer covering.
3. Crustaceans, centipedes and millipedes, spiders and their relatives, and insects.
4. Trap them in webs, leap out and grab them, and hide and jump out.
5. The soft shells are new shells forming after the blue crab molts to accommodate growth.

REINFORCEMENT/RETEACHING

Monitor students' responses to the Section Review questions. If students appear to have difficulty with any of the questions, reteach the appropriate section material.

CLOSURE

▶ *Review and Reinforcement Guide*
Students may now complete Section 2–1 of the *Review and Reinforcement Guide.*

2-2 Insects: The Most Numerous Arthropods

TEACHING STRATEGY 2-2

FOCUS/MOTIVATION

Present a graph to help students better conceptualize the vast number of insect species inhabiting the Earth. Because about three fourths of all animal species are insects, you could draw a circle on the chalkboard, and with colored chalk, section off a wedge comprising one fourth of the circle to represent all animals other than insects. The remaining three fourths of the circle represents the 750,000 or more species of insects. Point out that at least 275,000 of the insect species are beetles.

Media and Technology

Have students examine the relationships between insects and their environment through the use of the Interactive Videodisc called Insects: Little Giants of the Earth. At this point, students can discover how insects behave, communicate, and defend themselves. Students can also

---

Guide for Reading

Focus on these questions as you read.

▶ What are the characteristics of insects?
▶ How do insects develop, behave, and defend themselves?

Figure 2–12 Compare this grasshopper with the diagram of the structure of a grasshopper. How many legs does an insect have? ❶

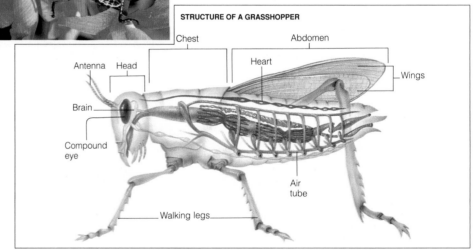

STRUCTURE OF A GRASSHOPPER

2–2 Insects: The Most Numerous Arthropods

Perhaps you have noticed that one group of arthropods mentioned in the previous section has not yet been discussed. This group, of course, is the insects. It would be hard to overlook insects for long. After all, there are more kinds of insects than there are all other animal species combined. In fact, it has been estimated that there may be as many as 300 million insects for every person on Earth!

Insect Structure

Insects are described as having a body that is divided into three parts—a head, a chest, and an abdomen—and that has three pairs (6) of legs attached to the chest part. In a grasshopper, which is a typical insect, the three pairs of legs are not identical. In order for a grasshopper to jump, one pair of legs must be larger than the other two pairs. Which pair of legs do you think is larger: the front pair or the back pair? ❷

If you look closely at the head of a grasshopper, you will see five eyes peering back at you. Three of these eyes are located on the front of the grasshopper's head and are called simple eyes. Simple eyes can detect only light and dark. The remaining two

---

explore the important contributions that insects have made to the environment.

CONTENT DEVELOPMENT

Direct the discussion on insect anatomy by asking questions similar to the following:

• **How is an insect's body different from that of an arachnid?** (An insect's body has three sections, but an arachnid's body has two sections.)

• **What are the names of the three main sections of a grasshopper's body?** (The sections are the head, thorax, and abdomen.)

• **How many pairs of legs does an insect have?** (An insect has three pairs of legs.)

• **What structures are located on a grasshopper's head?** (Three simple eyes and a pair of compound eyes.)

Point out that a pair of antennae are also on the head. The antennae contain an insect's organs for smelling.

eyes, which are found on each side of the grasshopper's head, are called compound eyes. Compound eyes contain many lenses. Although compound eyes can distinguish some colors, they are best at detecting movement. This ability is important to an animal that is hunted by other animals on a daily basis.

Like most insects, a grasshopper has wings—two pairs of wings, in fact. Although these wings are best suited for short distances, some types of grasshoppers can fly great distances in search of food. Insect flight varies from the gentle fluttering motion of a butterfly to the speedy movement of a hawkmoth, which can fly as fast as 50 kilometers per hour!

Like all insects, a grasshopper does not have a well-developed system for getting oxygen into its body and removing waste gases from its body. What a grasshopper does have is a system of tubes that carries oxygen through the exoskeleton and into its body.

Male grasshoppers produce sperm, and female grasshoppers produce eggs. Like most insects, a male grasshopper deposits sperm inside a female grasshopper during reproduction.

Growth and Development of Insects

Insects spend a great deal of time eating. As a result, they grow rapidly. And like other arthropods, insects must shed their exoskeletons as they grow. During the growth process, insects pass through several stages of development. Some species of insects change their appearance completely as they pass through the different stages. This dramatic change in form is known as **metamorphosis** (meht-uh-MOR-fuh-sihs). The word metamorphosis comes from the Greek words meaning to transform. There are two types of metamorphosis: complete and incomplete.

During complete metamorphosis, insects such as butterflies, beetles, bees, and moths pass through a four-stage process. The first stage produces an egg. When an egg hatches, a **larva** (LAHR-vuh) emerges, completing the second stage. A caterpillar is the larva of an insect that will one day become a butterfly or a moth. Maggots are the larvae of flies, and grubs are the larvae of some types of beetles. A larva spends almost all its time eating.

Figure 2–13 *Insects such as the horsefly have compound eyes. Within the eyes are many lenses that enable the insect to detect the slightest movement of an object.*

Figure 2–14 *Insects, birds, and bats are the only organisms that can fly on their own. Like all insects, the painted beauty butterfly and the drone fly have two pairs of wings.*

C ■ 49

BACKGROUND INFORMATION
INSECT DIGESTION

The mouths and digestive systems vary among the species of insects. Mosquitoes have mouths that resemble needles so that a mosquito can push its mouth through an animal's skin and withdraw body fluids. The grasshopper's digestive tract is specialized so that it can eat plant tissue. Salivary glands moisten food to help it move through the gut. From the mouth, food passes through the esophagus into a crop, where it can be stored. Next, food moves into the gizzard, where chitinous teeth grind it up further. In the intestine, glands produce enzymes to digest food, and other structures absorb the digested foods. Undigested food leaves through the rectum.

In bees, the large abdomen has a swollen area of the gut called the honey crop. Here, a mixture of nectar and saliva is converted into the more easily digestible form we call honey.

GUIDED PRACTICE

▶ *Laboratory Manual*

Skills Development

Skills: Making observations, making identifications, applying concepts, drawing conclusions

At this point you may want to have students perform the Laboratory Investigation called The Grasshopper, in the *Laboratory Manual.* In the investigation students examine the external structure, movements, and behavior of a grasshopper.

CONTENT DEVELOPMENT

Explain that even though an insect does not have a circulatory system, it does have a heart and one long blood vessel along the top side of the body. The heart is a long tube closed at the posterior end. When it contracts, blood is forced forward into the blood vessel, where it is carried to the vicinity of the head. From there, the blood goes into the body cavity, where it bathes the internal organs. The blood reenters the heart through openings located in the abdominal segment.

● ● ● ● **Integration** ● ● ● ●

Use the information about eye structure to integrate concepts of light into your lesson.

Use the etymology of the term *metamorphosis* to integrate language arts into your science lesson.

MOSQUITO BITES

Although almost everyone has been bitten by a mosquito, students may not be aware that only the female mosquito bites. Male mosquitoes cause no trouble at all, flying from flower to flower collecting pollen. Females use the nutrients in blood to help them produce large numbers of eggs. To prevent blood from clotting as they drink it, the mosquitoes inject their saliva when they first pierce the skin. The saliva can carry disease-causing organisms such as the malaria-causing protozoan. The body's reaction to this saliva causes the itching and swelling that comes with mosquito bites.

HISTORICAL NOTE
THE MOSQUITO AND THE CANAL

In 1904, when work on the Panama Canal began, workers often came down with yellow fever, which is spread by mosquitoes. An earlier attempt to build a canal by the French had been abandoned partly because yellow fever and malaria had not been controlled. By the time the American-built Panama Canal was finished, Colonel William C. Gorgas had established an extensive sanitary program that effectively curtailed the mosquito population and the incidence of yellow fever. Gorgas had earlier helped to conquer yellow fever in Havana, Cuba, following the Spanish-American War. In Cuba, the Cuban physician Carlos Finlay and the American doctor Walter Reed proved that yellow fever was spread by mosquitoes.

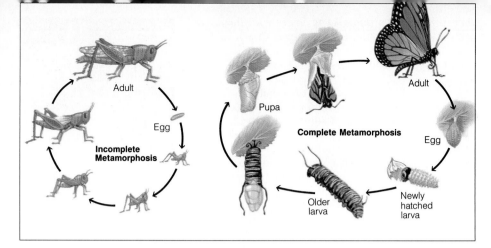

Figure 2–15 *The monarch butterfly (right) undergoes complete metamorphosis, whereas the grasshopper (left) undergoes incomplete metamorphosis. What is another name for the older larva of a monarch butterfly?* ❶

Figure 2–16 *The monarch butterfly begins life as an egg and then becomes a larva, also known as a caterpillar. During the pupa stage, the caterpillar wraps itself into a cocoon called a chrysalis. Finally, the butterfly emerges.*

Eventually, a larva enters the third stage, which produces a **pupa** (PYOO-puh; plural: pupae). Pupae are sometimes wrapped in a covering called a cocoon or chrysalis (KRIHS-uh-lihs). A cocoon is made of silk or other similar material. During this stage, remarkable changes take place, and an adult insect soon emerges. The adult signals the last stage of complete metamorphosis. An adult insect not only looks like a different animal, it also behaves differently.

During incomplete metamorphosis—which occurs in insects such as grasshoppers, termites, and dragonflies—young animals looking very much like the adults hatch from eggs. See Figure 2–15. These young animals do not have the organs of an adult, however, and they often do not have wings either. As the young animals grow, they keep molting (shedding

---

2–2 (continued)

FOCUS/MOTIVATION

Most students are familiar with the metamorphosis of frogs from tadpoles to adult stage. Show the class a picture of tadpoles and adult frogs or an illustration of the frog life cycle.

• **In what ways are the animals in this picture like those in Figures 2–15 and 2–16?** (Lead students to suggest that both frogs and butterflies undergo metamorphosis.)

• **What is meant by the term** *metamorphosis?* (Metamorphosis refers to a series of changes an organism undergoes as it develops from egg to adult.)

• **What are the stages of metamorphosis of a frog?** (Tadpoles hatch from eggs laid in water. The tadpoles gradually develop into adult frogs.)

• **Describe the metamorphosis of the monarch butterfly.** (A larva, or caterpillar, hatches from an egg. The larva forms a pupa from which the adult butterfly emerges.)

CONTENT DEVELOPMENT

Explain that not all insects go through a four-stage metamorphosis. In some wingless insects, such as springtails, the young are just like the adults except for size. Other insects undergo an incomplete metamorphosis consisting of three stages. An immature wingless nymph hatches from the egg; then undergoes several molts and gradually develops

their exoskeletons) and getting larger until they reach adult size. Along the way, the young animals acquire all the characteristics of an adult animal.

Insect Behavior

Most insects live solitary lives. In this way, they do not compete directly with other members of their species for available food. Male and female insects, however, do interact in order to reproduce. But before they do, they must attract or signal each other. This is done in a variety of ways, depending on the insects. For example, to attract a female, the male cicada (sih-KAY-duh) buzzes by, vibrating a special membrane in its abdomen. A male firefly attracts a female firefly, which in some species is called a glowworm, by turning the light-producing organ in its abdomen on and off. The female gypsy moth attracts a male by releasing extremely powerful chemicals called **pheromones** (FER-uh-mohnz). Pheromones cannot be detected by humans, but only a small amount of pheromones can attract a male gypsy moth from several kilometers downwind.

Other insects, known as social insects, cannot survive alone. These insects form colonies, or hives. Ants, termites, some wasp species, and bees are social insects. They survive as a society of individual insects that perform different jobs. As you can see in Figure 2–18, many of these colonies are highly organized.

One of the most fascinating examples of an insect society is a beehive. A beehive is a marvel of organization. Worker bees, which are all females, perform

Figure 2–17 *A male luna moth's feathery antennae can detect pheromones released by a female luna moth several kilometers away. What are pheromones?* ❷

Figure 2–18 *It may be hard to believe, but this mound was built by termites (left)! Termites, carpenter ants (center), and honeybees (right) are examples of social insects. How do social insects differ from most other insects?* ❸

C ■ 51

ECOLOGY NOTE
PESTICIDES

In some places, the widespread use of chemical pesticides "won the battles, but lost the war." For example, after years of spraying in the United States and Mexico, the numbers of cotton-eating pests such as the tobacco budworm and bollworm were greater than ever before because the pesticides had killed off their natural enemies. In addition, spraying transformed some minor pests into major ones; for example, the corn rootworm developed resistance to most pesticides and is now one of the worst pests of corn crops in the United States.

Activity Bank

Collecting Soil Critters, Activity Book, p. 155. This activity can be used for ESL and/or Cooperative Learning.

---

into a winged, sexually mature adult. Grasshoppers, termites, and dragonflies are examples of insects that undergo incomplete metamorphosis.

Media and Technology

Students can see how certain adaptations have made insects able to live in the near-constant dryness of the desert by viewing the Interactive Videodisc called On Dry Land: The Desert Biome. In subsequent chapters, students can continue their exploration of adaptation by examining other members of the animal kingdom that inhabit the desert biome.

REINFORCEMENT/RETEACHING

▶ *Activity Book*

For students who are having difficulty distinguishing the characteristics of arachnids and insects, assign the chapter activity called Insect or Arachnid? In the activity students identify similarities and differences in insects and arachnids and label specific organisms as insects or arachnids.

GUIDED PRACTICE

▶ *Laboratory Manual*

Skills Development

Skills: Making observations, recording data, making comparisons, applying concepts

At this point you may want students to complete the Laboratory Investigation called Observing Mealworms in the *Laboratory Manual*. In the investigation students observe the life cycle, anatomy, and behavioral characteristics of mealworms.

BACKGROUND INFORMATION
TERMITE SOCIETY

All species of termites are social, but there are fundamental differences between their species and those of bees. Among bees, only females contribute to activities of the colony because the males can function only during reproduction. In termite societies, however, all young nymphs resemble adults. Typically, some become workers. The workers are sterile and wingless. Others develop into soldiers or reproductive males and females. The soldiers have large, powerful jaws that they use to defend the nest.

The reproductive males and females of some species are winged, and at times they are produced in enormous numbers. The winged individuals then fly out in swarms to establish new colonies. On finding a new location for a nest, they land, shed their wings, and mate. Together the male (king) and female (queen) share the labor of establishing the new colony. Once the colony is established, the king and queen remain together for years and reproduce repeatedly.

ACTIVITY
WRITING

What Kind of Insect Is It?

There are many different kinds, or orders, of insects. Visit the library and look up the following information about insects: name of the order, some characteristics of the order, and two examples of insects found in that order. Arrange this information in the form of a chart.

Activity Bank

How Many Are Too Many?, p.159

Figure 2–19 *Insects defend themselves in a variety of ways. The tropical walking stick (top right) blends in with its surroundings so that it can hide from its predators. The bombardier beetle (bottom right) sprays a foul-smelling chemical. The peacock moth (bottom left) has eyespots that startle predators. How do you think the thorn bug (top left) defends itself?* ❶

all the tasks necessary for the survival of the hive. Worker bees supply the other members of the hive with food by making honey and the combs in which the honey is stored. They also feed the queen bee, whose only function is to produce an enormous number of eggs. In addition, worker bees clean and protect the hive. Male bees have only one function: to fertilize a queen's eggs.

Defense Mechanisms of Insects

Insects have many defense mechanisms that enable them to survive. Wasps and bees use stingers to defend themselves against enemies. Other insects are masters of camouflage (KAM-uh-flahj). This means that they can hide from their enemies by blending into their surroundings. For example, insects such as the stick grasshoppers resemble sticks and twigs. These insects survive because their bodies are not easily seen by their predators. See Figure 2–19.

2–2 (continued)

INDEPENDENT PRACTICE

▶ *Activity Book*

Students can reinforce their understanding of social insects by completing the chapter activity called Social Insects. In the activity students identify characteristics of specific insects.

GUIDED PRACTICE

▶ *Laboratory Manual*

Skills Development

Skills: Manipulative, making observations, comparing and contrasting, applying concepts

At this point you may want students to complete the Laboratory Investigation called Investigating an Ant Society in the *Laboratory Manual*. In the investigation students set up an ant colony and observe

PROBLEM Solving ？ ？ ？

Do You Want to Dance?

How do honeybees communicate information about the type, quality, direction, and distance of a food source to other members of the hive? The answer is that they do a little dance. Actually, they do two basic dances: a round dance and a waggle dance.

In the round dance, the honeybee scout that has found food circles first in one direction and then in the other up the honeycomb, over and over again. This dance tells the other honeybees that food is within 50 meters of the hive.

In the waggle dance, the honeybee scout that has found food runs straight up the honeycomb while waggling her abdomen, circles in one direction, runs straight again, and then circles in the other direction. This dance tells the other honeybees that food is more than 50 meters away from the hive. If food is located toward the sun, the honeybee scout will run straight up the honeycomb in the same direction as the sun. If food is located 30° to the right of the

sun, she will make a series of runs to the right of an imaginary vertical line on the honeycomb. If food is located 30° to the left of the sun, the same dance will be performed to the left of the imaginary line.

Use the four diagrams to answer the questions that follow.

1. Where is food located in Diagram C?

2. How far away is food in Diagram B: less than 50 meters or more than 50 meters?

3. Where is food located in Diagram A?

4. How far away is food in Diagram D: less than 50 meters or more than 50 meters?

Applying Facts

■ If you were a honeybee scout, how would you tell your hivemates that food is located more than 50 meters from the hive and 40° to the left of the sun?

■ Do honeybee scouts do waggle dances at night? Explain.

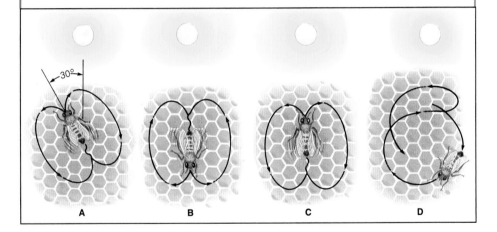

30°

A   B   C   D

PROBLEM SOLVING
DO YOU WANT TO DANCE?

This feature enables students to identify the way that one group of insects—the honeybees—communicate information about food sources to other members of the community.

1. Food is located directly toward the sun.

2. Food is located more than 50 meters from the hive.

3. Food is located 30 degrees to the right of the sun, more than 50 meters from the hive.

4. Food is less than 50 meters from the hive in Diagram D.

A honeybee scout would do the waggle dance and run 40 degrees to the left of the sun.

Honeybee scouts do not do the waggle dance at night because the sun is not available as a reference point.

the physical and behavioral characteristics of members of the society.

CONTENT DEVELOPMENT

Bees and wasps are commonly recognized as insects that use stingers to protect themselves. When they sting someone, they inject a poisonous secretion. Some flies and true bugs are also venomous, but they use their mouthparts rather than a stinger to inject their poison. Some insects defend themselves by spraying a foul-smelling secretion. The bombardier beetle is an example of an insect that sprays such a secretion. Some insects like the grasshopper have protective coloring or shapes that enable them to blend in with their surroundings. Other insects, such as some moths, have color patterns that startle or confuse their enemies. Still other insects protect themselves by mimicry; that is, these insects have color patterns that resemble those of another insect that the predator might avoid. For example, the wasp beetle has a black-and-yellow banded body that resembles that of a certain stinging wasp.

2-3 Echinoderms: The "Spiny-Skinned" Animals

MULTICULTURAL OPPORTUNITY 2-3

To help students understand the structure of the echinoderms, have them build a model of a starfish. They should find ways to illustrate the key characteristics of the echinoderms in their models.

ESL STRATEGY 2-3

Have students compare three invertebrates—a starfish, a crab, and a flatworm—explaining how they are similar. Point out that comparisons frequently use words such as *like, just as, too,* and *as well.*

Have students make models or drawings or bring in different types of echinoderms to demonstrate their distinguishing features to the class.

2-2 (continued)

INDEPENDENT PRACTICE

Section Review 2-2

1. Insects are arthropods that have a chest, head, and abdomen and have three pairs of legs attached to the chest part.
2. Change in appearance during development through adulthood. Egg, larva, pupa, adult.
3. Social insects live in colonies or hives in which individuals perform different tasks.
4. Defense mechanisms include stingers, camouflage, sprays, and patterns that startle or confuse predators.
5. Pheromone use would be safer to the environment because pheromones occur naturally, as opposed to many pesticides that are synthetic chemical compounds.

REINFORCING/RETEACHING

Review students' responses to the Section Review questions. Based on students' answers, reteach any material that may still be unclear.

2-2 Section Review

1. What are the characteristics of insects?
2. What is metamorphosis? What are the four stages of metamorphosis in a butterfly?
3. How do social insects live?
4. How do insects defend themselves?

Connection—*Ecology*
5. Which method of insect control would be considered more environmentally safe: the use of pheromones or the use of chemicals known as pesticides? Explain.

Guide for Reading

Focus on this question as you read.

▶ What are the characteristics of echinoderms?

Figure 2-20 *Starfishes such as the ochre sea star use their tube feet to open mussels.*

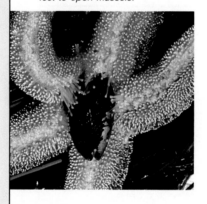

54 ■ C

2-3 Echinoderms: The "Spiny-Skinned" Animals

Have you ever seen a starfish? Look at the photograph in Figure 2–21. It shows some interesting examples of invertebrates that belong to the phylum Echinodermata (ee-kihg-noh-DER-muh-tuh). Members of this phylum are called echinoderms. Echinoderms include starfishes, sea lilies, feather stars, sea cucumbers, sea urchins, and sand dollars—to name just a few. The word echinoderm comes from the Greek words meaning spiny skin. As their name indicates, members of this phylum are spiny-skinned animals.

In addition to having a spiny skin, echinoderms have an internal skeleton, a five-part body, a water vascular system, and structures called tube feet. The internal skeleton of an echinoderm is made of bonylike plates of calcium that are bumpy or spiny. The **water vascular system** is a system of fluid-filled internal tubes that carry food and oxygen, remove wastes, and help echinoderms move. These tubes open to the outside through a strainerlike structure. This structure connects to other tubes, which eventually connect to the suction-cuplike **tube feet**. All echinoderms use their tube feet to "walk." Some echinoderms also use them to get food.

CLOSURE

▶ *Review and Reinforcement Guide*

Students may now complete Section 2-2 of the *Review and Reinforcement Guide.*

TEACHING STRATEGY 2-3

FOCUS/MOTIVATION

Display several pictures of starfishes and other echinoderms. Because many echinoderms are strikingly colorful, you may want to display the pictures on a bulletin board.

Explain that all these organisms are members of the phylum Echinodermata. Members of this phylum all have a spiny skin, an internal skeleton, a five-part body, a water vascular system, and tube feet. Then ask students the following questions:

• **Have would you describe the "skin" of a starfish?** (Many coarse spines cover the upper surface.)

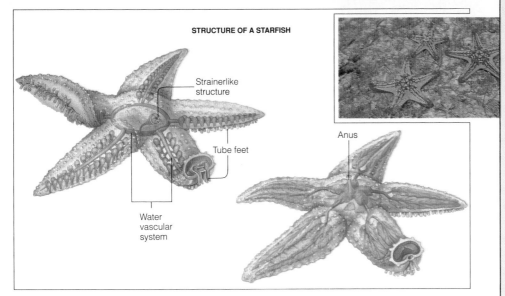

STRUCTURE OF A STARFISH

Strainerlike structure

Tube feet

Water vascular system

Anus

Starfishes

Although starfishes are not fish, most of them are shaped like stars. For this reason, starfishes are also called sea stars. Those that are star shaped have five or more arms, or rays, extending from a central body. On the underside of the arms are hundreds of tube feet that resemble tiny suction cups. These tube feet help the animal to move about and to obtain food. When a starfish passes over a clam, which is one of its favorite foods, the tiny tube feet grasp the clam's shell. The suction action of the hundreds of tube feet creates a tremendous force on the clam's shell. Eventually, the shell opens and the starfish enjoys a tasty meal.

People who harvest clams from the ocean bottom have long been at war with starfish that destroy the clam beds. In the past, starfishes captured near clam beds were cut into pieces and thrown back into the ocean. The people soon learned that, like flatworms and some crustaceans, starfishes have the ability to regenerate. By cutting them up, the people were guaranteeing that there would always be more and more starfishes—exactly the opposite of what they wanted to do.

Figure 2–21 *Echinoderm means spiny skin, which as you can see from this photograph of starfishes is an appropriate name for members of this phylum.*

Figure 2–22 *This sea star regenerated from a single arm.*

C ■ 55

can then attach to the surface by suction.
• **What are two functions of a starfish's tube feet?** (Locomotion and feeding.)
• **How does a starfish use its tube feet to open clam shells?** (The starfish spreads its arms over a clam's shell and attaches its tube feet to the two halves of the shell. It then uses the suction to gradually force open the shell.)

Explain that the starfish then extends its stomach through its mouth opening into the space between the two halves of the clam shell. Digestive juices produced by the starfish partially digest the clam within its own shell. The partially digested clam is taken into the stomach, which is then pulled back into the starfish.

• • • • **Integration** • • • •

Use the etymology of the term *echinoderm* to integrate language arts into your science lesson.

GUIDED PRACTICE

▶ *Laboratory Manual*

Skills Development

Skills: Making observations, drawing conclusions, applying concepts, making inferences

At this point you may want students to complete the Laboratory Investigation called The Starfish in the *Laboratory Manual*. In the activity students examine a preserved starfish and draw conclusions based on their observations.

• **What do you think is the function of the spines?** (They provide protection against predators.)
• **Where do you think the starfish's mouth is located?** (The mouth is in the middle of the central disc on the lower side.)
• **Do you think there are any sense organs on the starfish's body?** (Point out that an eyespot sensitive to light and a tentacle sensitive to touch are located at the end of each arm.)

• **How are starfishes like the other animals pictured?** (All are members of the same phylum.)

CONTENT DEVELOPMENT

Explain that a starfish has tube feet, a characteristic of all echinoderms. The tube feet are connected to a system of water canals inside the starfish's arms. Water can be forced from the canals into the tube feet, causing them to extend. The suckers of the extended tube feet

56 ■ C

Figure 2-23 *Some other examples of echinoderms include, from top to bottom, sea lilies, sand dollars, sea cucumbers, and sea urchins.*

Other Echinoderms

Members of the other groups of echinoderms vary widely in appearance. Sea lilies and feather stars, which are thought to be the most ancient of the echinoderms, look like flowers and stars with long, feathery arms. These echinoderms spend most of their time attached or clinging to the ocean bottom. They use their long, feathery arms to gather food from the surrounding water.

Sea cucumbers, as their name implies, resemble warty cucumbers, with a mouth at one end and an anus at the other. These animals are usually found lying on the bottom of the ocean. Sea cucumbers move along the ocean bottom by using the five rows of tube feet on their body surface to wiggle back and forth.

Sea urchins and sand dollars are round shaped and rayless. Sand dollars are flat, whereas sea urchins are dome shaped. Many sea urchins have long spines that they use for protection. In some of these sea urchins, poisonous sacs found at the tip of each spine can deliver painful stings.

2-3 Section Review

1. What are the characteristics of echinoderms?
2. List some examples of echinoderms.
3. What are two functions of a starfish's tube feet?
4. Why is cutting up a starfish and throwing it back into the ocean an ineffective way of reducing a population of starfishes?

Critical Thinking—*Making Comparisons*
5. What are some similarities between echinoderms and mollusks? What are some differences? Which group do you think is more complex? Explain.

CONNECTIONS

Insects in Flight ❶

Like a helicopter, a dragonfly can fly straight up or straight down. It can move to the right or to the left. It can glide forward or backward or simply hover in the air. And it can land on a lily pad in a pond without causing even the slightest ripple. It can reach speeds of 40 kilometers per hour and then stop on a dime.

As you can imagine, these amazing insects easily run circles around the best human-designed aircraft. For this reason, one group of researchers has been studying dragonflies to learn some of their secrets of *aerodynamics*. Aerodynamics is the study of the forces acting on an object (an airplane or a dragonfly) as it moves through the air.

One of the first goals of the researchers was to determine the lift that a dragonfly could produce. Lift is the force produced by the motion of a wing through the air. Lift is what gives an airplane the ability to climb into the air and hold itself upright during flight. Using a

tiny instrument that detects small forces, researchers measured the lift generated by several species of dragonflies. They discovered that dragonflies produce three times the lift for their mass. (The mass of a dragonfly is only about one seventh the mass of a dime!)

How can dragonflies perform this feat? Researchers discovered that dragonflies twist their wings on the downward stroke. This twisting action creates tiny whirlwinds on the top surfaces of the wings. This action moves air quickly over the wings' upper surfaces, lowering air pressure there and providing incredible lift.

By applying the aerodynamic principles of dragonfly flight to airplanes, scientists may soon be able to design and build more efficient airplanes. Of course, these future airplanes will never be able to bend or flex their wings as a dragonfly does. But they may be able to take off more easily, turn faster, and touch down on tiny landing fields.

C ■ 57

CONNECTIONS
INSECTS IN FLIGHT

Students may be fascinated by the study of the aerodynamics of the dragonfly. They may never have thought of insects in terms of the information that can be applied to manufactured materials. Discuss how engineers might apply the understanding of the lift of dragonflies to the design of airplanes. Also help students consider how the flight patterns of the dragonflies might be applied to aircraft.

If you are teaching thematically, you may want to use the Connections feature to reinforce the themes of scale and structure and systems and interactions.

Integration: Use the Connections feature to integrate the physical science concepts of aerodynamics into your lesson.

fishes develop from one that is thrown back in pieces.

5. Mollusks have soft bodies; echinoderms have spiny skins. Most mollusks live in the ocean; all echinoderms live in the ocean. Accept all logical answers. Although they have radial symmetry and seem simple, echinoderms are more closely related to vertebrates than are any of the other invertebrate phyla discussed in this textbook.

REINFORCEMENT/RETEACHING

Monitor students' responses to the Section Review questions. If students appear to have difficulty with any of the questions, reteach the appropriate section material.

CLOSURE

▶ *Review and Reinforcement Guide*
Students may now complete Section 2–3 of the *Review and Reinforcement Guide*.

Laboratory Investigation

INVESTIGATING ISOPOD ENVIRONMENTS

BEFORE THE LAB

1. Several days before the investigation, ask students to bring a shoe box to class. The same boxes can be used by different classes, and they can also be saved for use in future years.

2. One day before the investigation, ask students to collect at least 10 isopods. Students who have extra isopods can share them with students who do not have enough. To keep the isopods healthy until class time, students should be told to place several centimeters of slightly damp soil in their collecting jar and to keep the jar in a cool, dark place.

PRE-LAB DISCUSSION

Have students read the complete laboratory procedure.

• **What is the purpose of this investigation?** (To determine whether isopods prefer a wet or a dry environment.)

• **Based on what you already know about isopods, state a hypothesis.** (Accept any logical hypothesis. A possible hypothesis is, "If isopods are given a choice of a moist or a dry environment, they will choose the moist environment.")

• **How will you record your data for the three trials?** (Lead students to suggest that a data table is needed.)

Ask for suggestions in designing a data table. After a consensus is reached on a suitable format, place a table on the chalkboard as a guide for students to use in setting up their tables.

• **How will the class data be compiled?** (Each group's average results for each situation will be posted on the class data table.)

SAFETY TIPS

Remind students that isopods, like all living organisms, must be treated with care and in a humane manner.

Laboratory Investigation

Investigating Isopod Environments

Problem

What type of environment do isopods (pill bugs) prefer?

Materials *(per group)*

collecting jar	aluminum foil
10 isopods	paper towels
shoe box with a lid	masking tape

Procedure

1. With the collecting jar, gather some isopods. They are usually found under loose bricks or logs. Observe the characteristics of the isopods.

2. Line the inside of the shoe box with aluminum foil.

3. Place 2 paper towels side by side in the bottom of the shoe box. Tape them down. Place a strip of masking tape between the paper towels.

4. Moisten only the paper towel on the left side of the box.

5. Place 10 isopods on the masking tape. Then place the lid on the shoe box and leave the box undisturbed for 5 minutes.

6. During the 5-minute period, predict whether the isopods will prefer the moist paper towel or the dry paper towel.

7. After 5 minutes, open the lid and quickly count the number of isopods on the dry paper towel, on the moist paper towel, and on the masking tape. Record your observations in a data table.

8. Repeat steps 5 through 7 two more times. Be sure to place the isopods on the masking tape at the start of each trial. Record the results in the data table.

9. After you have completed the three trials, determine the average number of isopods found on the dry paper towel, on the moist paper towel, and on the masking tape. Record this information in the data table. Record your average results in a class data table on the chalkboard.

Observations

1. What characteristics of isopods did you observe?

2. Where were the isopods when you opened the lid of the box? Was your prediction correct?

3. Were there variables in the experiment that could have affected the outcome? If so, what were they?

4. What was the control in this experiment?

5. How did your results compare with the class results?

Analysis and Conclusions

1. Based on your observations of their characteristics, into which phylum of invertebrates would you classify isopods? Into which group within that phylum would you place isopods?

2. From the class results, what conclusions can you draw about the habitats preferred by isopods? Give reasons for your answers.

3. What was the purpose of the masking tape in the investigation?

4. Why did you perform the investigation three times?

5. **On Your Own** Design another investigation in which you test the following hypothesis: Isopods prefer dark environments to light environments. Be sure to include a variable and a control.

TEACHING STRATEGIES

1. Circulate about the room as students work to ensure they stay on task.

2. Insist that students enter data in their tables as it is collected rather than trying to remember it to record later. Assist any students who have difficulty computing their average results.

3. Check to see that all groups enter their data in the class data table.

DISCOVERY STRATEGIES

Discuss how the investigation relates to the chapter ideas by asking open questions similar to the following:

• **Is an isopod an insect? Explain your answer** (No. Insects have six legs; isopods have more than six legs—making comparisons.)

• **Why does the isopod prefer the moist environment?** (As a crustacean, it needs the moist environment to obtain oxygen through its gills—applying concepts.)

Summarizing Key Concepts

2–1 Arthropods: The "Joint-Footed" Animals

▲ Arthropods have an exoskeleton, a segmented body, and jointed appendages. An exoskeleton is a rigid outer covering.

▲ The process by which arthropods shed their exoskeleton as they grow is molting.

▲ Athropods include crustaceans, centipedes and millipedes, arachnids, and insects.

▲ A crustacean has a hard exoskeleton, two pairs of antennae, and mouthparts used for crushing and grinding food.

▲ Centipedes have one pair of legs in each body segment; millipedes have two pairs of legs in each body segment.

▲ The bodies of arachnids are divided into a head and chest part and an abdomen. Arachnids have four pairs of legs.

2–2 Insects: The Most Numerous Arthropods

▲ Insects have three body parts—head, chest, and abdomen—and three pairs of legs.

▲ The dramatic change in form an insect undergoes as it develops is called metamorphosis. There are two types of metamorphosis: complete and incomplete.

▲ During complete metamorphosis, insects go through a four-stage process: egg, larva, pupa, and adult.

▲ During incomplete metamorphosis, young insects looking very much like the adults hatch from eggs.

▲ Some species of insects give off extremely powerful chemicals called pheromones that attract either males or females.

2–3 Echinoderms: The "Spiny-Skinned" Animals

▲ Invertebrates with rough, spiny skin; an internal skeleton; a five-part body; a water vascular system; and tube feet are called echinoderms.

▲ Members of the phylum Echinodermata include starfishes, sea cucumbers, sea lilies, sea urchins, and sand dollars.

Reviewing Key Terms

Define each term in a complete sentence.

2–1 Arthropods: The "Joint-Footed" Animals	2–2 Insects: The Most Numerous Arthropods	2–3 Echinoderms: The "Spiny-Skinned" Animals
exoskeleton molting	metamorphosis larva pupa pheromone	water vascular system tube foot

• **Crustaceans have a pair of appendages attached to each of its segments. Two of the appendages are antennae. Based on your observations of the isopod's appendages, do you think it is a crustacean?** (Yes, because it has a pair of appendages on each segment of its body and it has antennae—making observations, applying facts.)

• **Could you predict what environment an ant would prefer based on your observations of the isopod?** (No. Although both animals are arthropods, they belong to different classes and have different environmental needs—making inferences, applying concepts.)

OBSERVATIONS

1. Students may observe the segmented body, number of appendages, and preference of the isopods.

2. They scattered, looking for a dark place; most would stay on the moist towel. Answers will vary, based on students' predictions.

3. If the lid of the shoe box was opened too many times, light could be a variable that could affect the outcome. Hidden variables could include heat and time of day.

4. Based on the choice of moisture as the variable, the control would be dryness. If dryness was considered the control, the variable would be moisture.

5. All groups should get similar results.

ANALYSIS AND CONCLUSIONS

1. Isopods are in the phylum Arthropoda and class Crustacea because they have segmented bodies, jointed legs, and exoskeletons.

2. Isopods prefer dark and moist environments. They need the moisture to obtain oxygen through their gills.

3. To separate the moist and dry towels.

4. To ensure more accurate results.

5. In the student-designed experiment, one shoe box with moist towels could be used. Because the variable is lightness or darkness, the lid should be cut in half so that one half of the box is covered and one half is not. Students would find that the isopods prefer dark environments.

GOING FURTHER: ENRICHMENT

Part 1

Have students try the experiment they designed in Analysis and Conclusions 5. If time does not permit this to be completed in class, interested students could perform their experiments as an out-of-class project.

Part 2

▮ If some students would like to maintain a culture of isopods, they can do so by placing them in a container such as a plastic shoe box filled with good garden soil. The soil should be kept slightly moist, and a piece of cut potato can be placed on the soil, cut side down, for food.

Chapter Review

Chapter Review

ALTERNATIVE ASSESSMENT

The *Prentice Hall Science* program includes a variety of testing components and methodologies. Aside from the Chapter Review questions, you may opt to use the Chapter Test or the Computer Test Bank Test in your *Test Book* for assessment of important facts and concepts. In addition, Performance-Based Tests are included in your *Test Book*. These Performance-Based Tests are designed to test science process skills, rather than factual content recall. Since they are not content dependent, Performance-Based Tests can be distributed after students complete a chapter or after they complete the entire textbook.

CONTENT REVIEW

Multiple Choice

1. b
2. c
3. b
4. b
5. c
6. b
7. d
8. b
9. d
10. a

True or False

1. T
2. F, Arthropod
3. F, gills
4. F, arachnids
5. T
6. T
7. T
8. T

Concept Mapping

Row 1: Crustaceans, Millipedes and Centipedes, Insects

CONCEPT MASTERY

1. Advantages are that the exoskeleton serves as protection against predators, serves as a point for muscle attachment, and prevents water loss from the body. Disadvantages include molting, which exposes the arthropod to danger from predators.

2. In complete metamorphosis, an insect passes through four stages of development: egg, larva, pupa, and adult. The adult looks and behaves differently than

Content Review

Multiple Choice

Choose the letter of the answer that best completes each statement.

1. Which is a characteristic of all arthropods?
   a. spiny skin   c. gills
   b. exoskeleton   d. backbone
2. Crustaceans obtain oxygen from the water through
   a. book lungs.
   b. water vascular systems.
   c. gills.
   d. air tubes.
3. Which is an example of a crustacean?
   a. sea urchin   c. grasshopper
   b. shrimp   d. scorpion
4. How many pairs of legs do millipedes have per body segment?
   a. 100   c. 1
   b. 2   d. 1000
5. In which stage of metamorphosis is an insect wrapped in a cocoon?
   a. egg   c. pupa
   b. larva   d. adult
6. Which invertebrate produces silk?
   a. lobster   c. mite
   b. spider   d. sand dollar
7. Which group includes animals that can fly?
   a. arachnids   c. crustaceans
   b. echinoderms   d. insects
8. Which is an example of a defense mechanism in insects?
   a. molting
   b. camouflage
   c. pheromone production
   d. metamorphosis
9. Starfishes belong to a group of invertebrates called
   a. crustaceans.   c. arachnids.
   b. arthropods.   d. echinoderms.
10. Which group of invertebrates have tube feet?
    a. echinoderms   c. millipedes
    b. crustaceans   d. arachnids

True or False

If the statement is true, write ''true.'' If it is false, change the underlined word or words to make the statement true.

1. Arthropods have a rigid outer covering called an <u>exoskeleton</u>.
2. <u>Crustacean</u> means joint footed.
3. Crabs have <u>book lungs</u>.
4. Spiders are <u>insects</u>.
5. Mites are <u>arachnids</u>.
6. Insects have <u>three</u> pairs of legs.
7. A caterpillar is an example of a <u>larva</u>.
8. <u>Starfishes</u> have a five-part body.

Concept Mapping

Complete the following concept map for Section 2–1. Refer to pages C6–C7 to construct a concept map for the entire chapter.

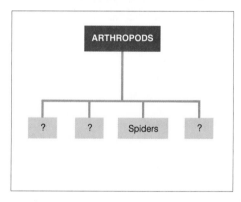

the insect does in other stages. In incomplete metamorphosis, the young insect that hatches from the egg looks very much like the adult. The young insect does not have all the needed organs and often does not have wings. As it grows, it acquires all the characteristics of the adult.

3. Crustaceans have gills, through which they obtain oxygen from water. Most spiders obtain oxygen through book lungs, which remove oxygen from passing air. Some spiders have a system of air tubes connected to the outside of the spider's body. Insects have a system of tubes that carry oxygen through the exoskeleton into the body.

4. Arthropods have jointed appendages that may include antennae, walking legs, wings, and claws.

5. Millipedes are arthropods with many legs. They have two pair of appendages on each segment and are plant-eating, shy creatures. Like millipedes, centipedes have many legs, but they have only one pair of legs per segment. They are carnivores with claws. Millipedes do not have claws. Both centipedes and milli-

Concept Mastery

Discuss each of the following in a brief paragraph.

1. What are some advantages of having an exoskeleton? What are some disadvantages?
2. Compare complete and incomplete metamorphosis.
3. Describe the different respiratory organs that are used by arthropods.
4. Describe the different types of arthropod appendages.
5. Compare millipedes and centipedes.
6. What is the function of tube feet?
7. How do starfishes move?
8. Why are arthropods the most numerous phylum of animals?
9. What role do pheromones have in the lives of insects?
10. Explain why arthropods undergo molting.
11. Describe the functions that a worker bee, a queen bee, and a male bee have in the hive.

Critical Thinking and Problem Solving

Use the skills you have developed in this chapter to complete each of the following.

1. **Making charts** Construct a chart in which you list the groups in the phyla Arthropoda and Echinodermata, the major characteristics of the group, and three animals from each group.
2. **Classifying objects** Your friend said he found a dead insect with two body parts and eight legs. Is this possible? Explain.
3. **Making generalizations** In what ways are insects beneficial to humans?
4. **Applying concepts** The makers of horror movies invent gigantic insects that terrorize human beings. Why is it impossible for such insects to exist?
5. **Relating concepts** Insects are often described as the most successful group of animals. What characteristics of insects could account for this description?
6. **Applying technology** Pesticides are chemicals used to kill harmful insects. Describe some advantages and disadvantages of pesticide use.
7. **Using the writing process** Observe an insect such as a bee or an ant for 15 minutes. Then write a short story describing what it would be like to be one of these animals.

pedes have exoskeletons that are not waterproof, and they are usually found in damp places.

6. Tube feet are used for moving and for getting food.

7. Starfishes move by tube feet.

8. They have evolved on Earth for over 600 million years and have developed characteristics, such as jointed legs, segmented bodies, and exoskeletons, that have made them successful.

9. Some insects use pheromones to attract a partner for mating.

10. Arthropods molt because their exoskeletons do not grow with the animal and must be shed to allow for growth.

11. Worker bees supply food by making honey and combs to store it in; they also clean and protect the hive. Queen bees lay eggs. Male bees fertilize the eggs.

CRITICAL THINKING AND PROBLEM SOLVING

1. Charts should be consistent with the information in the chapter.

2. The dead creature could not be an insect because insects have three body parts and six legs. It could be an arachnid, which has two body parts and eight legs.

3. Answers will vary, but may include these: Some insects eat other organisms that are harmful to crops and annoying or harmful to people; some insects pollinate flowers.

4. The exoskeleton needed to support such a large arthropod would have to be very thick and heavy. The arthropod would collapse under its own weight.

5. There are many species of insects; insects are found throughout the world; and insects are extremely diverse.

6. Pesticides can be used to protect crops from insects. Although they kill the undesirable insects, they can also kill other wildlife, including birds; they can seep into the water supply and cause health problems for wildlife and humans.

7. Students short stories should include physical and behavioral characteristics of the insect that they are imagining to be.

KEEPING A PORTFOLIO

You might want to assign some of the Concept Mastery and Critical Thinking and Problem Solving questions as homework and have students include their responses to unassigned questions in their portfolio. Students should be encouraged to include both the question and the answer in their portfolio.

ISSUES IN SCIENCE

The following issue can be used as a springboard for discussion or given as a writing assignment:

Millions of dollars are spent every year to control insect pests with chemical pesticides. Many people are opposed to this practice because it is costly and because the pesticides can also be harmful to beneficial insects, other animals, and humans. In some cases, the insects can become resistant to the pesticides. Others argue that if pesticides were not used to control insects, they would soon take over the Earth. Crops would be destroyed, resulting in widespread starvation, and diseases spread by insects would cause unbelievable misery and death. What is your opinion and why?

Chapter 3 FISHES AND AMPHIBIANS

SECTION	HANDS-ON ACTIVITIES
3–1 What Is a Vertebrate? pages C64–C66 Multicultural Opportunity 3–1, p. C64 ESL Strategy 3–1, p. C64	**Student Edition** ACTIVITY (Doing): Tunicates, p. C66 **Teacher Edition** Observing Vertebrates, p. C62d
3–2 Fishes pages C67–C76 Multicultural Opportunity 3–2, p. C67 ESL Strategy 3–2, p. C67	**Student Edition** ACTIVITY (Doing): The Invasion of the Lamprey, p. C71 ACTIVITY (Doing): Observing a Fish, p. C72 LABORATORY INVESTIGATION: Designing an Aquatic Environment, p. C84 ACTIVITY BANK: To Float or Not to Float? p. C161 **Laboratory Manual** The Fish, p. C69 Observing Schooling Behavior of a Fish, p. C75 **Activity Book** ACTIVITY: Simulating a Floating Fish, p. C49
3–3 Amphibians pages C76–C83 Multicultural Opportunity 3–3, p. C76 ESL Strategy 3–3, p. C76	**Student Edition** ACTIVITY (Doing): A Frog-Jumping Contest, p. C79 **Teacher Edition** Observing Frogs and Toads, p. C62d **Laboratory Manual** The Frog, p. C79 **Activity Book** ACTIVITY BANK: A Frog of a Different Color, p. C167
Chapter Review pages C84–C87	

OUTSIDE TEACHER RESOURCES

Books
Arnold, Caroline. *Electric Fish*, Morrow.
Sattler, Helen Roney. *Sharks: The Super Fish*, Lothrop.
Simon, Hilda. *Frogs and Toads of the World*, Lippincott.
Zappler, Georg, and Lisbeth Zappler. *Amphibians as Pets*, Doubleday.

Audiovisuals
Biology of the Amphibians, filmstrip with cassette, SVE
Biology of the Bony Fishes, filmstrip with cassette, SVE
Frogs and How They Live, video, AIMS Media
Life Story of a Toad, video, EBE

Vertebrates, four sound filmstrips, National Geographic
What Is a Fish? video, EBE

OTHER ACTIVITIES	MEDIA AND TECHNOLOGY
Review and Reinforcement Guide Section 3–1, p. C39	**Video** Animals With Backbones (Supplemental) **English/Spanish Audiotapes** Section 3–1
Student Edition ACTIVITY (Reading): A Fish Story, p. C74 **Activity Book** ACTIVITY: Sharks! p. C55 ACTIVITY: Analyzing Adaptations for a Life in Water, p. C59 ACTIVITY: Analyzing Data About World Fisheries, p. C61 **Review and Reinforcement Guide** Section 3–2, p. C41	**Video/Videodisc** Seeing Sense Super Scents **Courseware** Life Zones in the Ocean (Supplemental) **English/Spanish Audiotapes** Section 3–2
Student Edition ACTIVITY (Reading): Jumping Frogs, p. C81 **Activity Book** ACTIVITY: Cracking a Coldblooded Code, p. C53 **Review and Reinforcement Guide** Section 3–3, p. C45	**Video** Amphibians (Supplemental) **English/Spanish Audiotapes** Section 3–3
Test Book Chapter Test, p. C49 Performance-Based Tests, p. C109	**Test Book** Computer Test Bank Test, p. C55

*All materials in the Chapter Planning Guide Grid are available as part of the Prentice Hall Science Learning System.

CHAPTER OVERVIEW

All vertebrates, or animals with a backbone, belong to the phylum Chordata. At some time during their development, all chordata have three important characteristics: a nerve chord, a notochord, and a throat with gill slits. There are eight groups of vertebrates within the phylum Chordata; of these eight groups, six are coldblooded and two are warmblooded. Coldblooded vertebrates include fishes, amphibians, and reptiles; warmblooded vertebrates include birds and mammals.

Fishes are water-dwelling vertebrates that are characterized by scales, fins, and throats with gill slits. Jawless fishes such as lampreys are the most primitive of all fishes, lacking jaws, scales, paired fins, and bones. Cartilaginous fishes such as sharks have skeletons made of cartilage and scales. Bony fishes such as trout are the most specialized of all fishes, having paired fins, skeletons made of bones, and swim bladders.

Amphibians are vertebrates that typically spend the early portion of their lives in water, and live the later portion of their lives on land as adults. The most familiar amphibians are frogs and toads. Other examples of amphibians include salamanders, newts, and legless amphibians.

3–1 WHAT IS A VERTEBRATE?

THEMATIC FOCUS

The purpose of this section is to introduce students to the general characteristics of a vertebrate. All organisms possessing a notochord during some stage of development belong to the phylum Chordata. The notochord is a flexible structure made of cartilage that helps to support an animal. In certain animals, this notochord develops into a backbone. Animals possessing a vertebral column are called vertebrates. Within the vertebrate subphylum, there are eight classes. Birds and mammals are two classes of warmblooded vertebrates. There are six classes of coldblooded vertebrates, including lampreys and hagfishes, cartilaginous fishes, bony fishes, amphibians, and reptiles. In this chapter students will study the characteristics of fishes and amphibians and learn that these classes share common characteristics.

The themes that can be focused on in this section are evolution and stability.

*Evolution: Fishes were the first vertebrates to have evolved. Because the primitive Earth was thought to be covered by oceans early in its development, life originated in the oceans and evolved to include organisms adapted for land survival.

Stability: The various life functions of vertebrates such as fishes and amphibians serve to maintain a stable internal environment for these vertebrates. Warmblooded vertebrates maintain body temperature internally as a result of chemical reactions that occur within cells. Coldblooded vertebrates do not produce much internal heat and must rely on their environment for the heat they need.

PERFORMANCE OBJECTIVES 3–1

1. Identify the characteristics of vertebrates.
2. Compare the differences between warmblooded and coldblooded vertebrates.

SCIENCE TERMS 3–1

vertebrate p. C64
gill p. C64
coldblooded p. C66
warmblooded p. C66

3–2 FISHES

THEMATIC FOCUS

This section will introduce students to the main characteristics of fishes and the main features of the three different groups of fishes. Students will discover that fishes are water-dwelling vertebrates that are characterized by scales, fins, and throats with gill slits.

There are three groups of fishes, the most primitive of the three being the jawless fishes. Jawless fishes lack paired fins and scales, and some are parasites of other fishes, attaching themselves to hosts and sucking out body fluids and sometimes even the internal organs of their hosts. The most common jawless fish is a lamprey.

Sharks, skates, and rays compose a group known as the cartilaginous fishes. Though they have jaws and fins, their skeletons are made entirely of cartilage.

The most numerous and widespread fishes are the bony fishes, whose skeletons are composed mainly of bone. An important structure of bony fishes is their swim bladder, a structure filled with air that helps to keep the fishes afloat at various underwater depths.

The themes that can be focused on in this section are unity and diversity, scale and structure, systems and interactions, and energy.

*Unity and diversity: Although various fishes are different in habits and appearance, they are all chordates. Chordates have a nerve cord, a notochord, and a throat with gill slits during some part of their lives.

*Scale and Structure: The body structures of fishes vary along with their life functions.

Systems and interactions: Fishes respond and interact with their environment in ways that help them to gather food, reproduce, and protect themselves.

Energy: Fishes obtain energy by eating plants or other animals.

PERFORMANCE OBJECTIVES 3–2

1. Name the three groups of fishes and their characteristics.
2. Identify the function of a swim bladder.
3. Describe reproduction by external fertilization.

SCIENCE TERMS 3-2

external fertilization p. C69
internal fertilization p. C69
swim bladder p. C73

3-3 AMPHIBIANS
THEMATIC FOCUS

This section will introduce students to different types of amphibians and their characteristics. Amphibians include such animals as frogs, toads, salamanders, and newts.

Students will learn that the term *amphibian,* which means "double life," explains that amphibians are adapted to both water and land, tending to be aquatic as larvae and terrestrial as adults.

Although in some parts of the world hunters tip their arrows with toxins secreted by frogs, students will discover that humans tend to have little interaction with amphibians. Nevertheless, amphibians play an important role in ecological systems, particularly as predators of insects.

The themes that can be focused on in this section are unity and diversity, scale and structure, patterns of change, systems and interactions, and energy.

*Unity and diversity: Even though amphibians such as frogs and salamanders can look very different from each other and have different habits, they are both chordates.

*Scale and structure: The body structures of amphibians such as frogs, toads, salamanders, and newts vary along with their life patterns.

*Patterns of change: Many amphibians undergo a series of dramatic changes in body form, which is known as metamorphosis.

Systems and interactions: Amphibians respond to and interact with their environment in ways that help them gather food, reproduce, and protect themselves.

Energy: Amphibians obtain energy by eating plants or other animals.

PERFORMANCE OBJECTIVES 3-3

1. Identify common characteristics of amphibians.
2. Compare various types of amphibians.
3. Describe the life cycle of a frog.
4. Define hibernation.

SCIENCE TERM 3-3

metamorphosis p. C79

Discovery *Learning*
TEACHER DEMONSTRATIONS MODELING

Observing Vertebrates

If possible, obtain a live representative of each vertebrate organism students will study in this chapter. Attempt to include a jawless fish, a cartilaginous fish, a bony fish, a frog or toad, and a salamander or newt. If live representatives are not available, gather pictures of each vertebrate. After students have had an opportunity to view each vertebrate, ask:
• **These organisms all look quite different, yet they have one important trait or characteristic in common. In what way are these organisms alike?** (Accept any sensible answer at this time. Lead students to discuss the internal skeletons of each organism, including the backbone.)
• **What are these animals with backbones called?** (These animals are vertebrates.)
• **What are some ways in which these animals are different?** (Accept all logical answers but focus attention on the different types of body coverings.)

Conclude by finding out how many students can determine the vertebrate group to which each specimen or pictured organism belongs.

Observing Frogs and Toads

If possible, obtain one or two live frogs and toads. Display the frogs in a glass aquarium containing a small amount of water and a rock or other object on which the frogs can climb out of the water. The toads can be displayed in a similar aquarium, but place loose soil and a small pan of water in it. Be sure to cover the tops of both containers.

If live animals cannot be obtained, display several pictures or wall charts that illustrate the distinctive features of both animals. After students have had an opportunity to view the living specimens or the pictures, call attention to some of the similarities and differences between frogs and toads by asking questions similar to these:
• **In what ways are the skins of frogs and toads different?** (Students should notice that a frog's skin appears smooth and thin, whereas the skin of the toad appears bumpy and thicker.)
• **In what ways are these animals alike?** (Accept all answers. Students will likely mention that the body shapes are similar and the hind legs are larger than the front legs.)
• **Which of these amphibians is likely to be found farthest away from water?** (Lead students to suggest that toads are better adapted for life on land.)

Mention that toads often bury themselves in moist soil to escape heat and enemies.

CHAPTER 3
Fishes and Amphibians

INTEGRATING SCIENCE

This life science chapter provides you with numerous opportunities to integrate other areas of science, as well as other disciplines, into your curriculum. Blue numbered annotations on the student page and integration notes on the teacher wraparound pages alert you to areas of possible integration.

In this chapter you can integrate language arts (pp. 64, 74, 77, 81), life science and evolution (pp. 64, 67, 74, 79), life science and homeostasis (p. 66), life science and classification (p. 70), physical science and electricity (p. 73), social studies (p. 76), and life science and viruses (p. 83).

SCIENCE, TECHNOLOGY, AND SOCIETY/COOPERATIVE LEARNING

Amphibians have survived on Earth for approximately 350 million years. Their biological adaptations have enabled them to endure life during a variety of environmental conditions. Today, however, dramatic drops in the number of amphibian species have been recorded on every continent. The cause for this decline reflects the deterioration of the Earth's environment.

Because of their unique biology, amphibians are good indicators of the Earth's environmental health, and the indications are not too good! Because they live part of their lives on land and part in water, amphibians are susceptible to both land- and water-based pollutants and contaminants. The ability to breathe through their semipermeable skin makes amphibians sensitive to air and water pollution and perhaps even to small increases in the amounts of ultraviolet radiation reaching Earth's surface due to the destruction of the ozone layer. Their main source of food—insects—guarantees them an abundant supply of things to eat. Insects, however, are the main target and common carriers of toxic pesticides used worldwide. The destruction of amphibian habitats on every continent is wiping out species before they can even

INTRODUCING CHAPTER 3

DISCOVERY LEARNING

▶ *Activity Book*

Begin teaching the chapter by using the Chapter 3 Discovery Activity from the *Activity Book*. Using this activity, students will explore how submerged fishes change depth.

USING THE TEXTBOOK

Have students observe the picture on page C62.

• **Describe the picture.** (The picture shows a male sea horse giving birth to live young.)

• **Do you think it is unusual that the male of a species such as the sea horse gives birth to live young?** (Accept all responses. Most students will probably think it is unusual.)

• **Why or why not?** (Responses will vary.

Fishes and Amphibians

Guide for Reading

After you read the following sections, you will be able to

3–1 What Is a Vertebrate?
- Describe the characteristics of vertebrates.
- Compare warmblooded and coldblooded vertebrates.

3–2 Fishes
- Identify the three groups of fishes and give an example of each.

3–3 Amphibians
- Describe the main characteristics of amphibians.
- Explain how metamorphosis occurs in frogs.

A sea horse is truly an unusual animal. It has the arching neck and head of a horse, the grasping tail of a monkey, and the color-changing power of a chameleon. It has eyes that move independently of each other, so that while one looks under water the other scans the surface. As if all this were not remarkable enough, sea horses have one more interesting feature: male sea horses, not female sea horses, give birth to baby sea horses!

If you look closely at the photograph on the opposite page, you can see the tails of a few baby sea horses sticking out of their father's pouch. A male sea horse bears the young. A female sea horse deposits eggs in a male sea horse's kangaroolike pouch, where they are fertilized and then cared for by the male sea horse. A few weeks later, the first baby sea horse is born, then another and another. The process continues until hundreds of tiny sea horses have emerged.

It may surprise you to learn that sea horses are actually fishes. Yes, fishes—complete with gills and fins. In the pages that follow, you will discover more about other fascinating fishes. You will also learn about the distant relatives of fishes: the amphibians.

Journal *Activity*

You and Your World Visit a supermarket and find out what kinds of fishes are available as food. In your journal, make a list of these fishes. Then choose one fish from your list and find a recipe for preparing it. Copy the recipe into your journal. Then, with the help of an adult, try it out.

A male sea horse giving birth to live young

Students may respond that it is unusual because the female of a species typically gives birth to its young.)
- **Can you think of any other organisms in which the male of the species gives birth to young?** (Answers will vary.)

Point out that there are other fishes in nature that, for example, are born male and then develop into females later in life. There are also a number of fishes that begin their lives as females and then change into males. Tell students that in this chapter they will study both fishes and amphibians.
- **What is a fish?** (Lead students to suggest that a fish is a vertebrate animal that is adapted for living its life under water.)
- **Can you name several examples of fishes?** (Examples will vary. As students name various fishes, tell them the type of fishes they have named, such as bony, cartilaginous, or jawless.)

---

be identified. The development of amphibian eggs laid in aquatic environments is being hindered by acidic precipitation and runoff of pesticides and fertilizers used in agriculture. Human consumption of frog legs, a delicacy in many parts of the world, has further reduced the numbers of some species of frogs.

Amphibians are among the most abundant creatures on Earth. But if the environmental health of the Earth continues to decline, there will continue to be a reduction in the worldwide population of amphibians. This reduction would only further contribute to the environmental imbalance on Earth, as food chains in every ecosystem would be severely affected. Can the cycle be stopped?

Cooperative learning: Using preassigned groups or randomly selected teams, have groups complete one of the following assignments.
- Construct a chart that identifies biological adaptations of amphibians, the environmental hazards faced because of these adaptations, and the role of humans in creating these hazards.
- Provide groups with information on the discovery of Australia's gastric brooding frog and its unexpected extinction in the early 1980s. (If time permits, you may want groups to perform this research themselves.) Have groups analyze the life cycle, habitat, and niche of the gastric brooding frog as they hypothesize about the cause of its extinction.

See Cooperative Learning in the *Teacher's Desk Reference.*

JOURNAL ACTIVITY

You may want to use the Journal Activity as the basis of a class discussion. After students create a list of the fishes found in a supermarket, have a volunteer write the names of the different kinds of fishes on the chalkboard. Then have other volunteers relate various things they may know about each of the fishes, such as where it is found naturally, its size, structure, or distinctive markings, and the countries that might rely on it as a significant food source. Students should be instructed to keep their Journal Activity in their portfolio.

3-1 What Is a Vertebrate?

MULTICULTURAL OPPORTUNITY 3-1

Some people view animals in a type of hierarchy—with some animals being more important than others. Is it possible that one reason for the widespread destruction of tropical rain forests and the loss of hundreds of species may be a lack of understanding of some forms of animal life?

ESL STRATEGY 3-1

Have students choose two animals—one vertebrate and one invertebrate—and outline or illustrate and label the particular characteristics of each of their chosen animals. Also have them compare the functions of each skeleton. Ask an English-speaking student to act as a tutor in helping LEP students prepare their work.

Guide for Reading

Focus on these questions as you read.

- *What are the main characteristics of vertebrates?*
- *What are three characteristics of chordates?*

3–1 What Is a Vertebrate?

What do trout, frogs, snakes, turtles, robins, bats, and humans have in common? The answer to this question is that all these animals are **vertebrates. A vertebrate is an animal that has a backbone, or a vertebral column.** The vertebral column of a vertebrate is important because it protects the spinal cord, which runs through the center of the backbone. The spinal cord is the connection between a vertebrate's well-developed brain and the nerves that carry information to and from every part of its body.

The vertebral column makes up part of a vertebrate's endoskeleton, or internal skeleton. (Remember, the prefix *endo-* means inner.) The endoskeleton provides support and helps to give shape to the body of a vertebrate. One important advantage of an endoskeleton is that it is made of living tissue, so it grows as the animal grows. This is quite unlike the exoskeleton of an arthropod, which is made of nonliving material and has to be shed as the animal grows.

All vertebrates belong to the phylum Chordata (kor-DAT-uh). Members of the phylum Chordata are known as chordates. **At some time during their lives, all chordates have three important characteristics: a nerve cord, a notochord, and a throat with gill slits.** The nerve cord is a hollow tube located near the animal's back. Just beneath the nerve cord is the notochord. The notochord is a long, flexible supporting rod that runs through part of the animal's body. In most vertebrates, the notochord is replaced by the vertebral column. The gill slits are paired structures located in the throat (or pharynx) region that connect the throat cavity with the outside environment. Water easily flows over the **gills,** allowing oxygen to pass into the blood vessels in the gills and carbon dioxide to pass out into the water. Gills are feathery structures through which water-dwelling animals, such as fishes, breathe.

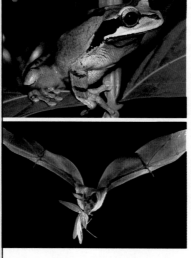

Figure 3–1 *Vertebrates are animals that have a vertebral column. Examples of vertebrates include the frog and the bat. To what phylum do vertebrates belong?* (1)

Figure 3–2 *In addition to illustrating one hypothesis about the evolutionary relationships among vertebrate groups, this* (2) *phylogenetic tree also shows approximately when certain characteristics occurred. When did four limbs appear?* (2)

TEACHING STRATEGY 3-1

FOCUS/MOTIVATION

Obtain a whole skeleton of a coldblooded vertebrate such as a fish and a warmblooded vertebrate such as a chicken. (If actual skeletons are not available, a picture of each will suffice.) Display the skeletons or pictures, and ask:

- **What animals are these skeletons from?** (A fish and a chicken.)
- **How are these skeletons different?** (They each have different shapes and numbers of bones.)
- **How are these skeletons similar?** (Both skeletons have a backbone that connects the other bones.)

Have students note the backbone of each animal. Point out that a fish and a chicken are both vertebrates because they are animals with backbones.

CONTENT DEVELOPMENT

Emphasize the idea that even though all vertebrates have backbones, they differ in many other ways. For example, some vertebrates are coldblooded, such as fishes, whereas others are warmblooded, such as chickens.

- **What characteristics do a fish and a sea turtle have in common?** (Both are coldblooded vertebrates adapted to life in water.)
- **What characteristics do a chicken and a human have in common?** (Both are warmblooded vertebrates adapted to life on land.)
- **What effect does environment have on the body temperature of a coldblooded vertebrate?** (The body temperature of a coldblooded animal may change when the temperature of its environment changes.)
- **What effect does environment have on the body temperature of a warmblooded vertebrate?** (The body temper-

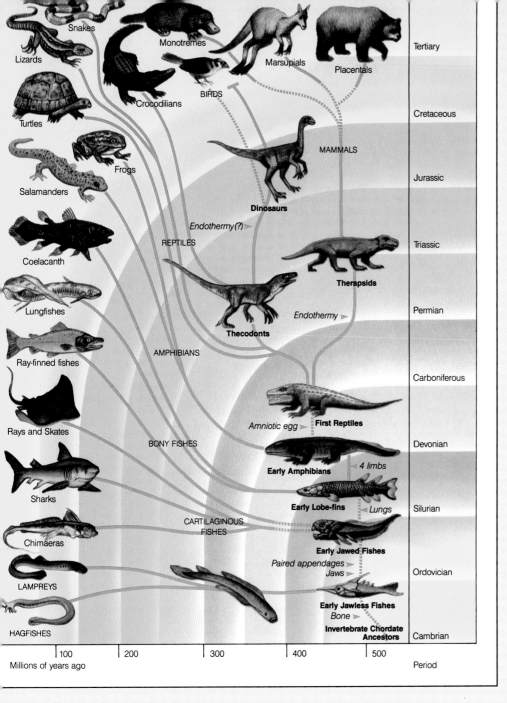

Snakes

Monotremes

Marsupials

Placentals

Tertiary

Lizards

BIRDS

Crocodilians

Cretaceous

Turtles

MAMMALS

Frogs

Jurassic

Salamanders

Dinosaurs

Endothermy (?)

REPTILES

Triassic

Coelacanth

Therapsids

Lungfishes

Endothermy

Permian

Thecodonts

Ray-finned fishes

AMPHIBIANS

Carboniferous

Rays and Skates

BONY FISHES

Amniotic egg   First Reptiles

Devonian

4 limbs

Sharks

Early Amphibians

Early Lobe-fins   Lungs

Silurian

CARTILAGINOUS
FISHES

Chimaeras

Early Jawed Fishes

Paired appendages

Jaws

Ordovician

LAMPREYS

Early Jawless Fishes

Bone

HAGFISHES

Invertebrate Chordate
Ancestors

Cambrian

| 100 | 200 | 300 | 400 | 500 |

Millions of years ago

Period

FACTS AND FIGURES

CHORDATE FOSSIL EVIDENCE

Scientists have no direct fossil evidence linking living vertebrates with lower chordates. The reason for this is that animals with no hard body parts do not fossilize well.

ature of a warmblooded vertebrate remains quite constant.)

● ● ● ● **Integration** ● ● ● ●

Use the introduction of the term *endoskeleton* to integrate language arts concepts into your lesson.

Use Figure 3–2 to integrate concepts of evolution into your lesson.

GUIDED PRACTICE

Skills Development

Skill: Making comparisons

Working in groups, ask students to develop a list containing the names of 20 different animals. Then have each group determine whether their listed animals are warmblooded or coldblooded. Volunteers might share their results with the class.

CONTENT DEVELOPMENT

Point out that there are eight classes of vertebrates within the phylum Chordata and that of the eight classes, two are warmblooded and six are coldblooded. Remind students that the behavior of a coldblooded animal changes when the temperature of its environment changes and then ask volunteers to describe any situations in which they have seen the behavior of an animal change because of temperature, such as a turtle sunning on a log during a cool, clear afternoon.

ACTIVITY
DOING
TUNICATES

Skills: Relating facts, applying concepts

Students should discover that tunicates are members of the subphylum Urochordata. A tunic, or covering, surrounds the animal's body. These animals have a notochord, which is a flexible rod running along the back of the animal. Tunicates also have a tubular nerve cord during their larval stage of development. The nerve cord is lost when they become adults. All tunicates are marine chordates, and some live in colonies. Examples of tunicates include sea squirts.

Students should determine that tunicates display three chordate characteristics during their lives—a nerve cord, a notochord, and a throat with gill slits.

3-1 (continued)

● ● ● ● **Integration** ● ● ● ●

Use the discussion of how warmblooded vertebrates regulate internal body temperature to integrate concepts of homeostasis into your lesson.

INDEPENDENT PRACTICE
Section Review 3-1
1. Vertebrates have an internal skeleton, or endoskeleton, and have a vertebral column made up of vertebrae, which protects the spinal cord.
2. At some time during their lives, all chordates have a nerve cord, a notochord, and a throat with gill slits.
3. Warmblooded animals have internal

Figure 3-3 *Vertebrates may be either coldblooded or warmblooded. The iguana, which is a reptile, is a coldblooded vertebrate. The polar bear, which is a mammal, is a warmblooded vertebrate.*

ACTIVITY

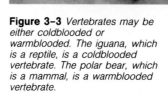

Tunicates

Tunicates are members of the phylum Chordata. Using reference materials in the library, find out about these animals. Present your findings to the class in an oral report. Include a drawing of a tunicate.

Why are these animals classified as chordates?

66 ■ C

There are eight groups of vertebrates within the phylum Chordata. Of the eight groups, six are **coldblooded** and two are **warmblooded.** Coldblooded animals (more correctly called ectotherms), such as fishes, amphibians, and reptiles, do not produce much internal heat. Thus they must rely on their environment for the heat they need. Warmblooded animals (endotherms), such as birds and mammals, maintain their body temperatures internally as a result of all the chemical reactions that occur within their cells. In other words, coldblooded animals have body temperatures that change somewhat with the temperature of their surroundings; warmblooded animals maintain a constant body temperature.

3-1 Section Review

1. What are the main characteristics of vertebrates?
2. List three characteristics of chordates.
3. Compare a coldblooded animal and a warmblooded animal.
4. What are gills?

Critical Thinking—*Applying Concepts*
5. Are you warmblooded or coldblooded? Explain your answer.

mechanisms that help to keep their internal body temperature constant despite changes in the external environment. Coldblooded animals also must keep their internal body temperature fairly constant and achieve this by adjusting their behavior during times of hot and cold weather. During cold weather, a coldblooded animal might bask in the sun, or during hot weather, the animal might rest in the shade, for example.
4. Gills are feathery structures through

which water-dwelling animals obtain oxygen and breathe.
5. Warmblooded. Explanations may vary, but students might suggest that humans are warmblooded because they maintain a constant body temperature.

REINFORCEMENT/RETEACHING

Monitor students' responses to the Section Review questions. If students appear to have difficulty with any of the questions, review the appropriate material in the section.

3-2 Fishes

About 540 million years ago the first fishes appeared in the Earth's oceans. These fishes were strange-looking animals, indeed! They had no jaws, and their bodies were covered by bony plates instead of scales. And although they had fins, the fins were not like those of modern-day fishes. But these early fishes were the first animals to have vertebral columns. They were the first vertebrates to have evolved.

Despite these differences, there was something special about these animals—something that would group them with the many kinds of fishes that were to follow millions of years later. **Fishes are water-dwelling vertebrates that are characterized by scales, fins, and throats with gill slits.** It is important to note, however, that not all fishes have all these characteristics. For example, sturgeons, paddlefishes, and sea horses have no scales at all on most of their body. And although most fishes have fins, the fins vary greatly in structure and function. Some fishes have paired fins, whereas others have single fins. Some fishes use their fins to help them remain upright. Other fishes use their fins to help them steer and stop. The side-to-side movement of large tail fins helps most fishes to move through the water. However, all fishes have gill slits.

Guide for Reading

Focus on these questions as you read.

▶ *What are the main characteristics of fishes?*

▶ *What are the main features of the three groups of fishes?*

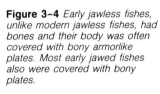

Figure 3-4 *Early jawless fishes, unlike modern jawless fishes, had bones and their body was often covered with bony armorlike plates. Most early jawed fishes also were covered with bony plates.*

3-2 Fishes

MULTICULTURAL OPPORTUNITY 3-2

The carp is admired in Japanese culture for its ability to fight its way upstream and its determination in overcoming obstacles. The carp symbolizes perseverance and courage. An old custom, still popular in Japan, requires a colorful carp-kite to be flown on May 5 (once called Boy's Day) from the home of every family with a son. The kite symbolizes the hope that a son will grow to be steadfast, like the carp, and that this character trait will bring him success in life.

Japanese culture also values *kingyo,* or the goldfish, for its graceful appearance and brilliant color. Goldfishes are bred as a hobby as well as for commercial use and are kept in elaborate glass receptacles or artificial ponds. Japan exports large quantities of goldfishes.

ESL STRATEGY 3-2

ESL students sometimes need practice in alphabetizing. Therefore, help them practice fish classification by having them place the three main groups of fishes in alphabetical order as headings on a chart. Then they should classify the following fishes, placing them in alphabetical order under the proper heading:

sharks electric eels perches trout African lungfishes hagfishes skates chimaeras flounders coelacanths sea horses lantern fishes remoras electric rays pancake sharks rays sawfishes lampreys mudskippers tuna

CLOSURE

▶ *Review and Reinforcement Guide*

At this point have students complete Section 3-1 in the *Review and Reinforcement Guide.*

TEACHING STRATEGY 3-2

FOCUS/MOTIVATION

Display a picture of the Earth taken from space. (Such pictures can be found in many earth science textbooks or in library books about planet Earth or space exploration.) Ask:

• **If you knew nothing about the Earth except for seeing this picture, how would you describe the planet?** (Answers may vary. Students may describe the planet as a swirling mass of water.)

• **Again, based only on this picture, what types of living things would you expect to find on this planet?** (Answers may vary. Students should make the connection that based on the planet's watery appearance, they would expect to find living things that live in water.)

● ● ● ● **Integration** ● ● ● ●

Use the description of the first fishes in the Earth's oceans to integrate concepts of evolution into your lesson.

FISH OR FISHES?

Although there is general confusion about the proper use of the words *fish* and *fishes,* the rule followed by biologists is quite simple. When referring to a single individual of a single species, use the word *fish.* When referring to many individuals of a single species, the proper plural is also *fish.* When referring to several individuals of more than one species, however, the proper plural is *fishes.*

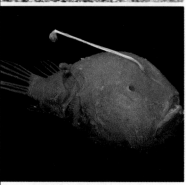

Figure 3–5 *Fishes have developed special structures that enable them to catch or eat a great variety of foods. The parrotfish (top) uses its "beak" to bite off chunks of coral. The oyster toadfish (center) relies on its ability to blend in with its surroundings to catch its prey. The seadevil (bottom) uses its bright lure to attract unsuspecting victims.*

As a group, fishes eat just about everything—from microscopic algae to worms to dead fish. The parrotfish even eats coral! Fishes have developed special structures that enable them to catch or eat the great variety of foods upon which they feed. Swordfishes are thought to slash their way through large groups of fishes and then return to devour the wounded or dead prey. Toad fishes rely on their ability to blend in with their surroundings to catch their prey. And angler fishes have wormlike lures that they dangle in front of their prey.

Like all vertebrates, fishes have a closed circulatory system. A closed circulatory system is one in which the blood is contained within blood vessels. In fishes, the blood travels through the blood vessels in a single loop—from the heart to the gills to the rest of the body and back to the heart. The excretory system of fishes consists of tubelike kidneys that filter nitrogen-containing wastes from the blood. Like many other water-dwelling animals, most fishes get rid of the nitrogen-containing wastes in the form of ammonia.

Fishes have a fairly well-developed nervous system. Almost all fishes have sense organs that collect information about their environment. Most fishes that are active in daylight have eyes with color vision almost as good as yours. Those fishes that are active at night or that live in murky water have large eyes with big pupils. Do you know what this adaptation enables them to do? ❶

Many fishes have keen senses of smell and taste. For example, sharks can detect the presence of one drop of blood in 115 liters of sea water! Although most fishes cannot hear sounds well, they can detect faint currents and vibrations in the water through a "distant-touch" system. As a fish moves, its distant-touch system responds to changes in the movement of the water, thus enabling the fish to detect prey or to avoid objects in its path.

3–2 (continued)

CONTENT DEVELOPMENT

Tell students that with more than two thirds of the Earth's surface covered by water, it is not surprising that the Earth looks like the "water planet." It is also not surprising that the most numerous group of vertebrate animals on Earth are aquatic.

CONTENT DEVELOPMENT

Have students observe Figure 3–5. Ask:

• **How are the fishes in these pictures similar?** (All the fishes are coldblooded vertebrates, adapted to underwater life, with scales, fins, gills, and special structures that enable them to catch or eat a great variety of foods.)

• **How are the fishes in these pictures different?** (The fishes are various sizes, shapes, and colors.)

Have students read the caption to Figure 3–5. Ask:

• **Do you think the parrotfish could survive in arctic waters?** (Lead students to suggest no. The parrotfish is adapted to warm, tropical waters and the presence of coral.)

Emphasize that each type of fish is suited to a particular environment. Conditions such as whether the water is fresh or salty and the temperature of the water will determine if a particular type of fish can survive there.

• **What do fishes found in the Great Salt Lake of Utah have in common with those found in the Pacific Ocean?** (They are both coldblooded vertebrates requiring a saltwater environment.)

• **Could a fish taken from the Great Salt Lake live in the Pacific Ocean?** (Its survival would be dependent on the water temperature of its new home.)

• **Could a fish native to Lake Michigan survive in the waters off the Florida coast?** (It is unlikely because the Florida waters would be warm salt water.)

ENRICHMENT

Various underwater creatures are native to a particular geographic location. For example, lobsters are found off the coast of New England, and shrimp are found off the coast of Louisiana. Have volunteers determine the underwater an-

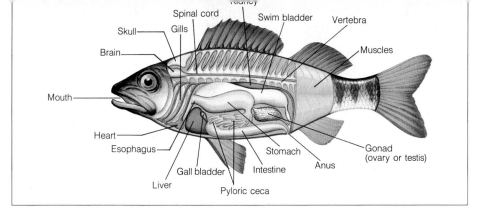

In most species of fishes, males and females are separate individuals. The males produce sperm, and the females produce eggs. There are, however, a number of fish species that are born males but develop into females. Others begin their lives as females and then change into males. Whatever the case, few fishes function as both a male and a female at the same time.

Of the many fishes that lay eggs, most have **external fertilization.** External fertilization is the process in which a sperm joins with an egg outside the body. Certain egg-laying fishes have **internal fertilization.** Internal fertilization is the process in which a sperm joins with an egg inside the body. Of the fishes that have internal fertilization, some—such as sharks and rays—lay fertilized eggs. In other fishes that have internal fertilization, the eggs develop inside the female's body. In this case, each developing fish receives food either directly from the female or indirectly from a yolk sac attached to its body. When all the food in its yolk sac is used up, the young fish is born.

Figure 3–6 *The internal organs of a typical bony fish are shown here. What structures enable fishes to breathe?* ❷

Figure 3–7 *Fishes have well-developed sense organs. For example, the eyes of the four-eyed fish, or quatro ojos, are each divided horizontally into two sections (right). The upper eyes are used to see above the water, whereas the lower eyes are used to see below the water. The "distant touch" system of the rainbow trout, which appears as a series of tiny dots in the pink stripe (left), detects movements in the water.*

BACKGROUND INFORMATION

SCALES

Fishes have many adaptations for living in water. One of these adaptations is scales. Slow-moving fishes that live near the bottom of water usually have large scales that serve as armor. Fishes that swim closer to the surface have small, lightweight scales that serve as protection.

fin on the bottom (where the body begins to taper) is called the anal fin. The dorsal and anal fins function to keep the fish from rolling. The pectoral and pelvic fins are the fins that are paired along the side of the body, and they are used for steering. The tail fin, also known as the caudal fin, provides thrust for movement.
• **What structures enable a fish to breathe in water?** (Fishes have gills for breathing.)

 Media and Technology

Students can get an explanation of the senses of sight and smell in animals such as fishes by viewing the Videos/Videodiscs entitled Seeing Sense and Super Scents. At this time students should begin their exploration by examining how and what animals see and smell. In subsequent lessons and chapters, students can continue their exploration of these senses in birds and mammals.

imals native to your geographic region and investigate how their presence has influenced local cuisine. Volunteers may share their findings with the class.

CONTENT DEVELOPMENT

Write this sentence on the chalkboard: Fishes are the vertebrates that are best adapted to life under water. Discuss some of these adaptations by having students observe Figure 3–6 and asking questions similar to the following:

• **How does the shape of a fish's body help adapt it for a life in water?** (The streamlined shape allows it to glide easily through water.)
• **In what way do fins enable a fish to move in water?** (Fins perform such functions as stabilization, steering, stopping, and propulsion.)
As students observe Figure 3–6, explain that the fins of a fish are supported by bones. The two fins on the top of the body of a fish are called dorsal fins. The

The parasitic sea lamprey is found in both fresh water and salt water. Though the sea lamprey is a notorious parasite, most other species of lampreys are small eellike creatures that feed on decayed organic material. Besides lampreys, the only other type of jawless fish is the hagfish. Like lampreys, hagfishes lack jaws, scales, and paired fins. Hagfishes live only in the sea as bottom-dwelling scavengers.

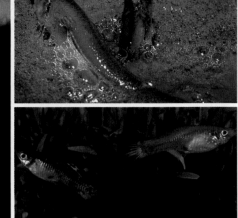

Figure 3–8 *Although all sharks have internal fertilization, the young of some sharks, such as the swell shark (left), develop outside the female's body. In grunions (top right), fertilization and development of young occur outside the female's body. In guppies (bottom right), fertilization and development of young occur inside the female's body.*

Many fishes (including some you can keep in an aquarium) have interesting mating behaviors. For example, male guppies dance in front of female guppies to get their attention. The bright red and blue body of a male three-spined stickleback serves to let female sticklebacks know where his nest is, as well as to warn other males to keep away.

The correct scientific classification of fishes is quite complicated. Thus, in this textbook fishes are ❶ placed into three main groups. These groups are the jawless fishes, the cartilaginous (KAHRT'l-aj-uh-nuhs) fishes, and the bony fishes.

Jawless Fishes

Jawless fishes are the most primitive of all fishes. They are so primitive, in fact, that in addition to lacking jaws, they also lack scales and paired fins. Something else makes jawless fishes unusual. To find out what it is, hold one of your ears between your fingers and move it back and forth a few times. How are you able to perform this action? The answer is that your ear contains a flexible material called cartilage. The entire skeleton of jawless fishes is made of cartilage. Jawless fishes are vertebrates that have no bone at all. (Later in this section you will read about another group of fishes that also contain no bone.)

70 ■ C

3-2 (continued)

CONTENT DEVELOPMENT

Emphasize the idea that during the process of external fertilization, the female fish releases her eggs directly into the water. The number of eggs released can be hundreds or thousands, depending on the type of fishes. Ask:

• **Does every egg released by the female develop into a fish?** (No, only the fertilized eggs will develop into fish.)

• **What happens to the unfertilized eggs?** (They are consumed by living things in the water.)

• **What might happen if every released egg was fertilized?** (The waters of the Earth would soon become overcrowded.)

Remind students that the correct classification of fishes is quite complicated. You might choose to discuss with them several reasons why the classification of fishes is so complicated.

● ● ● ● **Integration** ● ● ● ●

Use the introduction of the three main groups of fishes to integrate classification concepts into your lesson.

ENRICHMENT

▶ *Activity Book*

Students will be challenged by the Chapter 3 activity called Analyzing Data About World Fisheries. In this activity students will use data about the world-

wide production of fishes to determine what is happening to the fisheries of the world.

CONTENT DEVELOPMENT

Have students observe the lamprey shown in Figure 3–9. Ask:

• **How is a lamprey's mouth different from the mouths of other fishes?** (The lamprey has a suction-cuplike mouth without jaws.)

• **How is this mouth adapted to obtain food?** (Answers may vary. Call attention to the suckerlike mouth, which is used to attach the lamprey to a fish. Students should also note the lamprey's rasping teeth, which are used to drill into its prey.)

Mention that the lamprey is capable of sucking out a fish's internal organs as well as its blood and body fluids. These destructive predators may grow up to 1 meter in length. Because some students are often fascinated by the grotesque ap-

The only support the eellike bodies have is a noto-chord. As you might expect, jawless fishes are really flexible.

To see examples of the only two species of jaw-less fishes still alive—lampreys and hagfishes—look at Figure 3–9. Notice that a lamprey looks like an eel with a suction-cup mouth at one end. This suc-tion-cup mouth, which is surrounded by horny teeth, is extremely efficient. Using its mouth, a lamprey attaches itself to a fish such as a trout (or sometimes a whale or a porpoise) and scrapes away at the fish's skin with its teeth and rough tongue. Then it sucks up the tissues and other fluids of its victim.

The skin of a lamprey is covered with glands that release a slippery, sticky substance called mucus (MYOO-kuhs). This mucus is toxic, or poisonous, and it probably discourages larger fishes from eating lampreys.

The other jawless fishes—the hagfishes—are con-sidered the most primitive vertebrates alive today. The most obvious feature of a hagfish's wormlike body is the four to six short tentacles that surround its nostrils and mouth. The tentacles are used as organs of touch.

A hagfish feeds on dead or dying fishes by tear-ing out pieces of the fish with a tongue that has teethlike structures. If the fish is large, a hagfish will twist its own body into a knot so that it can thrust itself into the fish with extra power. In no time, the hagfish will be completely inside the prey fish. The ability to twist itself into a knot also enables a hag-fish to evade capture, especially when this action is accompanied by the release of a sticky, slimy mate-rial from pores located along the sides of its body. A single hagfish can release so much slime that if it is placed in a pail of sea water, it can turn the entire contents of the pail into slime.

Cartilaginous Fishes

When you think of sharks, you probably think of large fast-swimming vicious predators. Although this is true of some sharks, it is not true of most. For the most part, sharks prefer to be left alone.

Sharks—along with rays, skates, and two rare fishes called sawfishes and chimaeras (kigh-MIHR-uhz)—are cartilaginous fishes. Like jawless fishes,

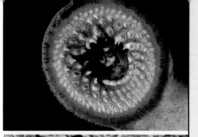

Figure 3–9 *Modern jawless fishes include only two species: the lampreys (top) and the hagfishes (bottom). In addition to being jawless, what are some other characteristics of jawless fishes?* ❶

ACTIVITY
DISCOVERING

The Invasion of the Lamprey

The completion of the St. Lawrence Seaway acciden-tally introduced lampreys into the Great Lakes. Using ref-erence materials in the li-brary, find out what effects lampreys have had on fishes already living there. Con-struct a chart that indicates what fishes lived in the Great Lakes before and after the arrival of lampreys.

■ What might have been done to prevent this from happening?

ACTIVITY
DISCOVERING
THE INVASION OF THE LAMPREY

Discovery Learning

Skills: Relating cause and effect, hypothesizing

Students should discover that lamprey movement into the Great Lakes caused a startling decline in the commercial catch of lake trout. For example, in Lake Mich-igan, the annual catch was reduced from 2,542,500 kilograms in 1945 to only 180 kilograms in 1954. Lampreys are now controlled by the use of selective poisons in their larval beds; as a result, the lake trout are making a comeback.

The various reasons suggested by stu-dents to prevent a lamprey invasion might include the installation of physical barriers for the lampreys.

pearance of a lamprey, obtain (if possi-ble) a preserved specimen from a biolog-ical supply company for students to examine.

Point out that all lampreys lay their eggs in freshwater streams. The young are only about 1 centimeter long when they burrow into mud at the bottom of a stream, remaining there for several years. After they mature, they move into open waters, where they begin to para-sitize fishes. The lamprey's reproduction and early development in freshwater streams have made it possible to control them to some degree by using selective poisons that kill the lamprey young while leaving other organisms unharmed. Ef-forts to curtail the lamprey population in the Great Lakes have been moderately successful.

GUIDED PRACTICE

Skills Development

Skills: Making comparisons, relating concepts

At this point have students complete the in-text Chapter 3 Laboratory Inves-tigation: Designing an Aquatic Environ-ment. In this investigation students will explore the type of environment best suited for guppies.

ACTIVITY
DOING

OBSERVING A FISH

Skills: *Making observations, relating concepts*

Materials: *fish, tray, hand lens*

Check the structures and labels of each diagram for accuracy. Students should note that most fishes have four gills enclosed in a gill chamber at each side of the head. Each gill is composed of two rows of fleshy filaments attached to a gill arch. A flap of bone called a gill cover protects the gills.

The class of fishes to which each fish belongs will vary, depending on the type of fishes. Most fishes, however, belong to the bony class of fishes.

3-2 (continued)

CONTENT DEVELOPMENT

Emphasize the idea that the skeletons of both jawless fishes and cartilaginous fishes are composed of cartilage. After students finish reading the material about cartilaginous fishes, have them compare the characteristics of jawless and cartilaginous fishes by asking questions similar to the following:

• **What are some traits shared by jawless fishes and cartilaginous fishes?** (Both are coldblooded vertebrates with skeletons composed of cartilage.)

• **How are jawless fishes and cartilaginous fishes different?** (Cartilaginous fishes have jaws.)

• **Why do you think jawless and cartilaginous fishes can survive without a strong skeleton made of bones?** (The body of the fish is supported by the water that surrounds it.)

Point out that one of the most recognizable cartilaginous fishes is a shark. Because sharks have a sleek appearance, students may sometimes assume that they have no scales. Be sure to emphasize that sharks do have scales but that they are different in structure and appearance from

Figure 3–10 *The southern stingray (left), big skate (bottom right), and sawfish (top right) are examples of cartilaginous fishes.*

ACTIVITY
DOING

Observing a Fish

1. Obtain a preserved fish from your teacher and place it in a tray.

2. Hold the fish in your hands and observe it. Note the size, shape, and color of the fish. Also note the number and location of the fins.

3. Draw a diagram of the fish and label as many structures as you can.

4. Locate the fish's gill cover. Lift it up and examine the gills with a hand lens.

To which group of fishes does your fish belong?

72 ■ C

cartilaginous fishes have skeletons made of cartilage. Most of them also have toothlike scales covering their bodies. The toothlike scales are the reason why the skin of a shark feels as rough as sandpaper. Most of the more than 2000 types of sharks have torpedo-shaped bodies, curved tails, and rounded snouts with a mouth underneath.

The most obvious feature of a shark is its teeth. At any one time, a fish-eating shark will have 3000 very long teeth arranged in six to twenty rows in its mouth. In most sharks, the first one or two rows of teeth are used for feeding. The remaining rows contain replacement teeth, with the newest teeth at the back. As a tooth in the front row breaks or is worn down, it falls out. When this happens, a replacement tooth moves forward in a kind of conveyor-belt system. In its lifetime, a single shark may go through more than 20,000 teeth! Not all sharks, however, have the long teeth characteristic of fish-eating sharks. Sharks that eat mollusks and crustaceans have flattened teeth that help them to crush the shells of their prey.

Unlike sharks, the bodies of skates and rays are as flat as pancakes. For this reason, skates and rays are sometimes called pancake sharks. These cartilaginous fishes have two large, broad fins that stick out from their sides. They beat these fins to move through the water, much as a bird beats its wings to fly through

the flat, overlapping scales of many bony fishes. A shark's body is covered with small, enamel-tipped, spinelike scales. These scales feel like sandpaper if they could be touched. The shark's teeth are similar in structure to the scales.

Also remind students that other fishes have different adaptations that help them successfully survive in an underwater environment.

● ● ● ● **Integration** ● ● ● ●

Use the discussion of specialized organs in fishes to integrate concepts of electricity into your lesson.

INDEPENDENT PRACTICE

▶ *Activity Book*

Students who need practice on the concept of cartilaginous fishes should complete the chapter activity Sharks! In this activity students will match descrip-

the air. Rays and skates often lie on the ocean bottom, where they hide by using their fins to cover their bodies with sand. When an unsuspecting fish or invertebrate comes near, the hidden skate or ray is ready to attack. Some rays have a poisonous spine at the end of their long, thin tail, which they use for defense rather than for catching prey. Other rays, appropriately called electric rays, have a specialized organ in their head that can discharge about 200 volts of electricity to stun and capture prey. Although 200 volts may not sound like a lot, you only need 120 volts to power almost everything in your home!

Bony Fishes

If you have ever eaten a flounder or a trout, you know why such fishes are called bony fishes. Their skeleton is made of hard bones, many of which are quite small and sharp. Some bony fishes, such as tunas, travel in groups called schools. Because of this schooling behavior, these fishes can be caught in large numbers at one time by people in fishing boats.

Although all bony fishes have paired fins, the shape of the paired fins varies considerably. Most bony fishes have fins supported by a number of long bones called rays. Thus these fishes are called ray-finned fishes. Perches and sea horses are two examples of ray-finned bony fishes. Other fishes have fins with fleshy bases supported by leglike bones. These fishes are known as lobe-finned fishes. Coelacanths (SEE-luh-kanths) are the only living species of lobe-finned bony fishes.

Another characteristic of bony fishes is that they have **swim bladders.** A swim bladder is a gas-filled sac that gives bony fishes buoyancy, or the ability to float in water. By inflating or deflating its swim bladder, a fish can float at different levels in the water.

There are many kinds of bony fishes, some of which have developed remarkable adaptations to life in water. For example, an electric eel can produce jolts of electricity up to 650 volts for use in defending itself or in stunning its prey. A remora uses its sucker to attach itself to sharks or other large fishes, feeding on bits of food they leave behind. Can you see why a remora is sometimes called a shark sucker?

Figure 3–11 *The sand tiger shark (top) shows one of the most noticeable characteristics of sharks: enormous numbers of teeth. The flattened body of the wobbegong, or carpet shark (center), and of the blue-spotted stingray (bottom) is an adaptation to life on the bottom of the ocean.*

Activity Bank

To Float or Not to Float?, p.161

C ■ 73

ANNOTATION KEY

Integration
① Physical Science: Electricity. See *Electricity and Magnetism,* Chapter 1.

FACTS AND AND FIGURES
THE MOVEMENT OF SHARKS

A shark will sink to the bottom of the ocean if it stops moving.

Dialogue and expression are the goals of this activity, so do not squelch any ideas that might be sincerely presented; however, you might tactfully attempt to clarify or correct any obvious misconceptions that are voiced concerning sharks.

tions and drawings and graph length and mass of various types of sharks.

GUIDED PRACTICE
Skills Development
Skill: Relating facts

Though most sharks are predators, the whale shark and the basking shark, which are the two largest sharks, feed only on small plants and animals. Though predatory sharks should not be taken lightly, only a few of them attack humans.

Give students an opportunity to express their opinions about whether or not they think sharks are vicious killers or merely predators in search of food. Consider using questions such as these to stimulate the discussion:
• **Do sharks have an undeserved bad reputation?**
• **Should all sharks be treated with extreme caution?**
• **Have some books and movies presented sharks in unrealistic ways?**

GUIDED PRACTICE
▶ *Laboratory Manual*
Skills Development
Skill: Making observations

At this point you may want to have students complete the Chapter 3 Laboratory Investigation in the *Laboratory Manual* called Observing Schooling Behavior of a Fish. In this investigation students will determine various interactions that can be observed in schooling fishes.

ACTIVITY
READING
A FISH STORY

FACTS AND FIGURES
MEDIEVAL FISH STORIES

In medieval times, it was widely believed that for every land animal there was a corresponding water animal. This is part of the reason for names such as catfish, sea horse, and dogfish. People were also thought to have corresponding water denizens, or inhabitants. As a result, skates and rays, often trimmed to enhance their resemblance to humans wearing conical hats, were once sold to the unsuspecting as genuine "sea bishops."

3-2 (continued)

CONTENT DEVELOPMENT

Remind students that the swim bladder is an important characteristic of bony fishes. The swim bladder is a cavity within the fish that can be filled with oxygen and nitrogen taken from the blood. By adjusting the amount of gas in its swim bladder, the bony fish adjusts itself to the water pressure present at various ocean depths.

• **What can a bony fish do because of its swim bladder?** (Float at any level in the water.)

• **How is a swim bladder similar to a life preserver?** (Both enable an object to float when inflated.)

• **What are some other objects humans use to float in water?** (Answers might include water wings, tire inner tubes, and rafts.)

Figure 3–12 *Bony fishes come in a wide variety of shapes and colors. The queen angelfish (bottom left) has a flattened, highly colorful body. The moray eel (top right) has a narrow, snakelike body. The body of the glass, or ghost, catfish (top left) is almost transparent except for its head and bones. Unlike its ancestors, which lived more than 70 million years ago, the present-day coelacanth (bottom right) still has its paired lobed fins.*

ACTIVITY
READING

A Fish Story

① You may want to read a wonderful novel about an old man and his struggle with nature as he pursues a large fish. The novel is entitled *The Old Man and the Sea,* and the author is Ernest Hemingway.

74 ■ C

Another fish that has developed an interesting adaptation to its surroundings is the flounder. All adult flounders are bottom-dwelling fishes. However, a flounder's eggs, which contain oil droplets, float at or near the surface of the water. When a young flounder begins its life, it does so as a normally shaped fish with one eye on each side of its head and a horizontal mouth. But as the young fish develops into an adult, one of its eyes moves to the other side of the head and the mouth twists. Because it does not have a swim bladder, the adult flounder eventually sinks to the ocean's bottom and lies permanently on one side—usually, the blind side. The fact that the flounder has its eyes and mouth on the same side of its body makes it easier for the flounder to see what is going on around it and to take in food. Lying on its side, a flounder is vulnerable to attack from its predators. But another adaptation— the ability to change the color of its body so that it matches the color of the ocean bottom—gives it protection from its predators. ②

Fishes that live in the depths of the ocean also have developed special adaptations. Lantern fishes, which live at depths of 300 to 700 meters, have light-emitting organs that attract prey. Other deep-sea fishes have huge eyes that help them to see better in the dark depths of the ocean.

GUIDED PRACTICE

Skills Development

Skill: Identifying patterns

Tell students that fishes are classified in various ways by ichthyologists, or scientists who study fishes. The system of classification presented in the textbook (jawless fishes, cartilaginous fishes, and bony fishes) represents the most generally used system. Have students suggest other ways in which fishes could be grouped.

CONTENT DEVELOPMENT

Emphasize the idea that most animals have certain adaptations for survival. Some adaptations help the animal live in a particular environment. Other adaptations help the animal protect itself from attackers. Ask:

• **Is the lungfish's ability to bury itself in the mud necessary for its defense or for its survival within its environment?** (For survival in an environment depleted of water.)

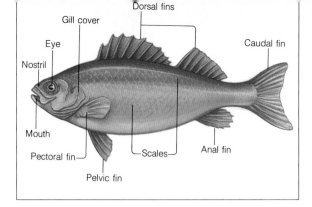

Dorsal fins

Gill cover

Eye

Nostril

Caudal fin

Mouth

Pectoral fin

Scales

Anal fin

Pelvic fin

Figure 3–13 *Although this diagram shows the different types of fins that may be present in fishes, not every kind of fish possesses all these fins.*

Still other fishes have developed adaptations that allow them to come out of water and spend some time on land. For example, mudskippers spend a lot of time using their fins to walk or "skip" on land at low tide. During these periods, a mudskipper can breathe air through its skin, as well as exchange oxygen for carbon dioxide in its mouth and throat. Another type of fish that can live on land for a short time is the African lungfish. When the swamp in which it lives dries up, an African lungfish burrows into the soft mud and becomes inactive until the rains come. When water is again available, the lungfish emerges.

Figure 3–14 *Bony fishes have developed many interesting adaptations to their life in water. The electric eel can discharge small amounts of electricity to protect itself from predators (bottom right). The remora, or sharksucker, has a suctionlike disk that it uses to "hitch a ride" on a shark (top right). The long, graceful rays of the lionfish contain poisonous spines, which help keep its predators away (left).*

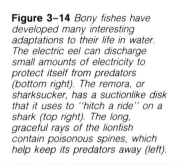
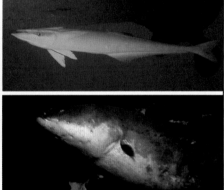

C ■ 75

INTEGRATION
BIOLOGY

Each particular fish has a built-in defense system. For example, the balloonfish can increase its size by suddenly taking in water. By doing so, the balloonfish becomes too large and cumbersome for its attacker. Have interested students investigate the ways in which other types of fishes fend off their attackers. Results can be compiled in written form or shared with the class.

INDEPENDENT PRACTICE

▶ *Activity Book*

Students who need practice on the concept of adaptations should complete the chapter activity Analyzing Adaptations for a Life in Water. In this activity students will examine several fish adaptations that enable fishes to perform life processes.

ENRICHMENT

Explain to students that the overall form and body plan of fishes tell much about the medium in which they live. For example, most fishes have streamlined bodies with a large proportion of skeletal muscle suited to forward motion. This is important because the density of water, which is about 800 times denser than air, makes movement through water difficult. Most fishes also have several pairs of fins, which are used to stabilize the body as well as propel and guide it through the water.

• **Is a flounder's ability to change color necessary for its defense or for its survival within its environment?** (Defense.)

● ● ● ● Integration ● ● ● ●

Use the description of the flounder's ability to change color to integrate concepts of evolution into your lesson.

REINFORCEMENT/RETEACHING

Be sure students understand the idea that in order to survive, a fish must be well suited to its environment. Have students determine whether the following statements are true or false:

• **Any fish can survive in any type of environment.** (False.)
• **Each fish requires a certain type of environment.** (True.)
• **Every fish has a particular water temperature range in which it can survive.** (True.)
• **All fishes can survive in the ocean.** (False.)

3-3 Amphibians

As you discuss metamorphosis, explain that the Greek prefix *meta-* implies changing, altering, or doing over and that *morphe* means form or shape. Thus, *metamorphosis* means a transformation—a change in form, structure, or substance.

Have students think about some of the changes, both physical and emotional, that they have gone through during their lives. Point out the analogy between these changes and the concept of metamorphosis.

ESL STRATEGY 3-3

Point out that the term *amphibian* can also be used to describe objects and plants. Ask students to explain what an amphibian airplane, amphibian tank, or amphibian plant might be capable of doing. Pair ESL students with English-speaking partners and have them make notes in order to be prepared to answer orally.

Enrich students' understanding of the words *metamorphosis* and *hibernation* by explaining that they are taken from Latin and Greek terms meaning "change of form" and "wintry."

3-2 (continued)

INDEPENDENT PRACTICE

Section Review 3-2

1. Fishes are water-dwelling vertebrates that are characterized by scales, fins, and gills.
2. Jawless fishes: lamprey; cartilaginous fishes: shark; bony fishes: trout. Examples may vary.
3. External fertilization is the process in which a sperm joins with an egg outside the body. Internal fertilization is the process in which the sperm joins with an egg inside the body.
4. A swim bladder allows a fish to float at various ocean depths by letting gases in or out of the bladder.
5. Answers may vary. Students might conclude that the reproductive rate of

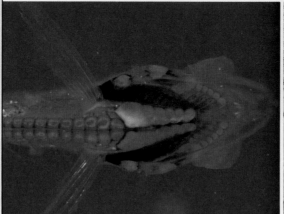

Figure 3–15 *Some bony fishes, such as the lantern fish, have light organs that enable them to live in the dark depths of the ocean. Other bony fishes, such as the African lungfish, have developed adaptations that allow them to live on land for short periods of time.*

3-2 Section Review

1. What are the main characteristics of fishes?
2. List the three groups of fishes and give an example of each.
3. Compare internal and external fertilization.
4. What is the function of a swim bladder?

Connection—*Ecology*

5. How would the Atlantic salmon be affected if its freshwater streams became badly polluted?

Guide for Reading

Focus on these questions as you read.

▶ What are the main characteristics of amphibians?

▶ What are some examples of amphibians?

76 ■ C

3-3 Amphibians

The forests of Colombia in South America are home to the Choco Indian tribe. There, the Indians continue a centuries-old tradition of hunting deer, monkeys, and even jaguars with poisoned arrows. In order to do so, the Indians must first capture a number of special kinds of frogs that live in the area. The Indians roast the frogs over a fire so that the poison drips from the skin. The poison is collected in pots and then smeared onto the tips of the arrows. The poison is so powerful that 0.00001 gram is enough to kill a person! So little poison is needed on the tip of an arrow that a 2.5-centimeter frog can

---

the salmon would decrease because salmon use freshwater streams as reproductive areas. Polluted streams would adversely affect the reproductive capacity of the fishes or the survival rate of eggs or newborns.

REINFORCEMENT/RETEACHING

Review students' responses to the Section Review questions. Reteach any material that is still unclear, based on students' responses.

CLOSURE

▶ *Review and Reinforcement Guide*

Students may now complete Section 3–2 in the *Review and Reinforcement Guide.*

TEACHING STRATEGY 3-3

FOCUS/MOTIVATION

Have students observe Figure 3–20. Point out that some people are unsure of

produce enough to cover 50 arrows. Appropriately, this frog is known as the arrow-poison frog.

Arrow-poison frogs are members of the second group of coldblooded vertebrates: the amphibians. Amphibians first appeared on Earth about 360 million years ago. They are thought to have evolved from lobe-finned bony fishes that had lungs—fishes similar to a modern coelacanth.

The word amphibian means double life. And most amphibians do live a double life. **Amphibians are vertebrates that are fishlike and that breathe through gills when immature. They live on land and breathe through lungs and moist skin as adults. Their skin also contains many glands, and their bodies lack scales and claws.** Naturally, there are exceptions. Some amphibians spend their entire lives on land. Others live their entire lives in water. But it is safe to say that most amphibians live in water for the first part of their lives and on land in moist areas as adults.

Why must most amphibians live in moist areas? One reason is that their eggs lack hard outer shells. If not deposited in water, such eggs would dry out. Another reason why adult amphibians cannot stray too far from a moist area is that in addition to breathing through lungs, they also breathe through

Figure 3–16 *There are three main groups of amphibians: frogs and toads, salamanders and newts, and legless amphibians. The red-eyed tree frog (left) can climb trees as well as hop. The red-bellied newt (top right) keeps its tail throughout its life. The burrowing caecilian (bottom right) preys on small animals it meets as it tunnels through the ground. What are the main characteristics of all amphibians?* ❶

the similarities and differences between frogs and toads. Ask:
• **What physical traits do frogs and toads share?** (Both have the same basic body structure; both have two short front legs and two long back legs; both have two eyes, a wide mouth, and a long tongue.)
• **What characteristics make toads different from frogs?** (Toads usually have drier skin than frogs do and are covered by warts. Frogs generally look smoother and sleeker than toads.)

• **Have you ever seen a frog or a toad?** (Yes.)
• **Where have you seen a frog or a toad?** (Answers may vary but should include land and water environments.)

Tell students that frogs and toads are classified as amphibians. These animals spend part of their lives on land and part in water. Therefore, it is possible to observe a frog or toad in either surrounding.

their skin. And in order to do so, the skin must remain moist. If the skin dries out, most amphibians will suffocate.

The circulatory system of adult amphibians forms a double loop. One loop transports oxygen-poor and oxygen-rich blood back and forth between the heart and the lungs. The other loop transports

Figure 3–17 *The major internal organs of a frog are shown in these two diagrams. Which organs enable the frog to breathe?* ❶

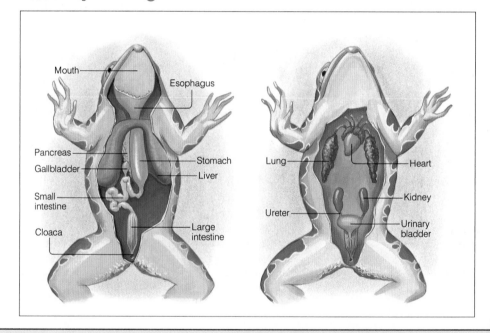

3-3 (continued)

CONTENT DEVELOPMENT

Emphasize the fact that though most amphibians can live on land as adults, they must still live in the vicinity of moisture.

• **Why must some amphibians return to water to reproduce?** (These amphibians lay unshelled eggs, which would dry out if laid on land.)

• **In addition to using lungs to breathe, how else do amphibians take in oxygen?** (Amphibians also breathe through their skin.)

• **What will happen if the skin of an amphibian dries out?** (The amphibian will not be able to breathe through its skin, and as a result, it will die.)

• **Why are amphibians not usually found very far from water or moist areas?** (Students should conclude that amphibians need water for reproduction and to keep their skin moist for gas exchange.)

● ● ● ● **Integration** ● ● ● ●

Use Figure 3–18 to integrate concepts of evolution into your lesson.

GUIDED PRACTICE

Skills Development
Skill: Hypothesizing

Have students observe Figure 3–18. Point out that the amphibians from about 270 million years ago, perhaps like those in the picture, became extinct soon after. About 245 million years ago, all that remained of the amphibians were three orders of small amphibians.

• **Why do you think the appearance of** land plants was important to the development of amphibians? (By the time amphibians were developed enough to live on land, the existing plants were important sources of food.)

• **What was another important food source for amphibians?** (Insects.)

• **What events do you think may have contributed to the extinction of large amphibians?** (Accept reasonable responses. Students might hypothesize that global changes in climate caused many of

oxygen-rich and oxygen-poor blood between the heart and the rest of the body. Tadpoles, or the young of certain amphibians, have a single-loop circulatory system, as do bony fishes. In this type of system, blood travels from the heart to the gills to the rest of the body and back to the heart.

Amphibians have two oval-shaped kidneys that filter wastes from the blood. The nitrogen-containing wastes are in the form of urine, which is then transported by tubes out of the body.

The nervous system and the sense organs are well-developed in amphibians. Large eyes, which bulge out from the sides of the head, provide sharp vision. A transparent membrane protects the eyes from drying out while the animal is on land and from being damaged while the animal is under water.

Many amphibians reproduce by external fertilization. In frogs, for example, a female releases eggs that are then fertilized by a male. After a sticky, transparent jelly forms around the fertilized eggs, the eggs become attached to underwater plants. In a few weeks the eggs hatch into tadpoles, or polliwogs. A tadpole has gills to breathe under water and feeds on plants. Eventually, the process of **metamorphosis** (meht-uh-MOR-fuh-sihs), or the series of dramatic changes in body form in an amphibian's life cycle, begins. During this process, the tadpole undergoes remarkable changes. It loses its tail and develops two

Figure 3–18 *This diagram shows what amphibians may have looked like 270 million years ago. What characteristics do they have in common with modern amphibians?* ❷

ACTIVITY
DOING

A Frog Jumping Contest

1. Find a flat surface that measures 1.5 m x 1.5 m. Draw four concentric circles with diameters of 30 cm, 60 cm, 90 cm, and 120 cm on the surface.

2. Place a frog in the middle of the innermost circle. Measure how far the frog jumps. Record the distance.

3. Repeat step 2 two more times. Find the average distance your frog jumps.

How far did your frog jump? Which legs do frogs use for jumping—the front legs or the hind legs? To which group of vertebrates do frogs belong? Explain your answer.

C ■ 79

the low, swampy, amphibian habitats to disappear.)

• **Why do you think that the three orders of small amphibians were able to survive?** (Accept reasonable answers. Students should realize that these organisms must have adapted to the changes in climate and developed other adaptations to ensure their survival. Also, the amphibians must have been successful in finding empty niches in their environment.)

REINFORCEMENT/RETEACHING

▶ *Activity Book*

Students who need practice on the concept of identifying amphibians should complete the chapter activity Cracking a Coldblooded Code. In this activity students will decode different symbols to discover the names of various coldblooded vertebrates.

CONTENT DEVELOPMENT

Emphasize to students that an important characteristic of amphibians is that they undergo a metamorphosis from a larval stage to an adult stage.

• **What is meant by the term *metamorphosis?*** (A change in form).

• **What type of metamorphosis do frogs undergo?** (They change from tadpoles into frogs.)

Remind students that there are many important differences between frogs and tadpoles and that many changes occur as a tadpole grows into an adult frog.

Frogs and toads have large tympanic membranes on either side of the head. These membranes are connected to the inner ear by a long bone called the stapes, which conducts sound. The frog's inner ear can distinguish among many different sounds.

pairs of legs. Its gills begin to disappear, and its lungs complete their development. The tadpole is now an adult frog, ready for life on land.

Not all amphibians lay eggs and have external fertilization. Of the many amphibians that have internal fertilization, some lay fertilized eggs. In others, the fertilized eggs develop inside the body of a female, where they receive their food directly from the female or indirectly from a yolk sac. Amphibians have varying ways of caring for their young. Some frogs carry their young in their mouth or in their stomach. Others have special structures on their back in which their young develop.

Figure 3–19 *Like most amphibians, frogs live in water for the first part of their life and on land in moist areas as an adult. Frog eggs are fertilized externally (top right) and generally develop in water (center left). Soon the fertilized eggs develop into young with tails (center right) and then hatch into tadpoles (bottom left). Gradually, the tadpoles grow limbs and begin to lose their tails (bottom right) as they develop into adults. What is this process called?* ❶

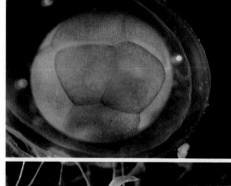

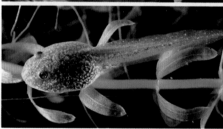

3–3 (continued)

CONTENT DEVELOPMENT

Emphasize the idea that during its life cycle, a frog passes through various stages of development. Have students observe Figure 3–19. Ask:
• **What is a young frog called?** (Tadpole or polliwog.)
• **How is a tadpole similar to a fish?** (Both have gills and a tail.)
• **Could a tadpole live on land?** (No.)
• **Why not?** (A tadpole has not developed lungs.)

GUIDED PRACTICE

Skills Development

Skill: Making observations

At certain times of the year—usually spring—frog egg masses or tadpoles can be collected from the shallow edges of ponds. At other times, they may be available from biological supply companies. If tadpoles can be obtained, have students work in small groups to design an experiment to test the effect of temperature on the development of tadpoles. Instruct groups that their designs must include the following:

1. A hypothesis to be tested
2. Means for establishing different temperature conditions
3. The number of tadpoles to be tested at each temperature
4. Procedures for making observations and recording data
5. Provisions for maintaining all tadpoles in good health throughout the experiment

If students undertake this activity, refer them to an appropriate reference on the care of tadpoles. After each group develops a plan, have the entire class discuss the merit of the individual components of each group's plan. If the plans are found to be adequate, the experiment may be attempted. As a suggested approach, students might keep one group of tadpoles at normal room temperature and another group at the same stage of development in a cool location, such as an unheated garage or a basement. Stress that throughout the course

Frogs and Toads

Have you ever wondered what happens to frogs and toads in the winter when the temperature falls? Frogs and toads, like all amphibians, are unable to move to warmer climates. They do, however, survive the cold. Frogs often bury themselves beneath the muddy floor of a lake during the winter. Toads dig through dry ground below the frost line. Then these amphibians go into a winter sleep called hibernation. During hibernation, all body activities slow down so that the animal can live on food stored in its body. The small amount of oxygen needed during hibernation passes through the amphibian's skin as it sleeps. Once warmer weather comes, the frog or toad awakens. If you live in the country, you can usually tell when this happens. The night suddenly becomes filled with the familiar peeps, squeaks, chirps, and grunts that male frogs and toads use to attract female mates.

Although frogs and toads appear similar in shape, you can discover one difference merely by touching

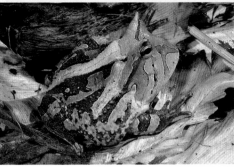

Figure 3–20 *Frogs and toads have developed adaptations that help them escape their predators. The tomato toad (bottom right) has glands behind its eyes that contain poison. The Amazon horned toad (bottom left) is almost invisible as it hides among dead leaves. The European tree frog (top) has long, muscular legs that enable it to quickly leap away from its enemies.*

C ■ 81

In preparation for this period of dormancy, the animals store energy in the bodies of fat located just above the kidneys and within the livers. Then they bury themselves in the mud beneath ponds and streams, slowing down their metabolism until the warm weather arrives in spring.

Amphibians that live in places where summers are hot and dry avoid drying out by estivating. When their water holes dry into mud flats, these tropical species dig deep burrows and coat the inside of their new homes with a mixture of mucus and dead skin. Inside these watertight burrows, some species can live in a kind of suspended animation until rain arrives again. A few species have been known to estivate for up to two years.

Have interested students investigate the ways in which other animals prepare for the winter season and have students report their findings to the class.

of the experiment, all tadpoles must be fed regularly and their water kept clean.

CONTENT DEVELOPMENT

Emphasize the idea that frogs and toads hibernate during the winter. Ask:
• **Is a frog a coldblooded or a warmblooded animal?** (Coldblooded.)
• **What effect does temperature have on a coldblooded animal?** (A coldblooded animal's body temperature will decrease as the temperature of its surroundings

decreases and will increase as the temperature of its surroundings increases.)
• **Why is hibernation a necessary behavior for frogs during cold winter months?** (By hibernating, a frog can conserve energy during periods when food is less abundant.)

ENRICHMENT

Many amphibians living in the temperate zone enter a dormant state, known as hibernation, during long, cold winters.

them. Frogs have a smooth, moist skin. Toads have skin that is drier and is usually covered with small wartlike bumps. In many toads, the bumps behind the eyes contain a poisonous liquid, which the toad releases when attacked. A great cane toad can squirt a jet of poison at an attacker almost a meter away. The attacking animal quickly becomes sick and may even die.

If there is one thing most people know about adult frogs and toads, it is that they are excellent jumpers. The main reason for this is that the hind legs of a frog or a toad are much larger than the front legs. It is these powerful hind legs that enable these animals to jump so well and that help them to escape from their enemies.

Salamanders and Newts

Salamanders and newts are amphibians that keep their tails throughout their lives. Like frogs and toads, these animals have two pairs of legs. But their hind legs are not as developed as those of a frog or a toad. Thus salamanders and newts are not able to jump.

Because they are amphibians, salamanders and newts must live in moist areas. One type of salamander, the mud puppy, lives in water all its life even though as an adult it has both lungs and gills. Like frogs and toads, salamanders and newts lay their eggs in water.

Figure 3–21 *Unlike most amphibians, salamanders and newts—such as the mud puppy (left) and red-spotted newt (right)—keep their tail throughout their life.*

82 ■ C

3-3 Section Review

1. What are the characteristics of amphibians?
2. List some examples of amphibians.
3. Explain how amphibians live double lives.
4. Compare a tadpole with an adult frog. List at least three differences between them.

Critical Thinking—*Relating Concepts*

5. Amphibians can lay as many as 200 eggs. Why do you think it is necessary for most amphibians to produce so many eggs?

CONNECTIONS

Can Toads Cause Warts? ❶

Have you ever been told that you can get warts by touching a toad? Contrary to superstition, touching the skin of a toad does not cause warts. Although a pair of large glands on the top of a toad's head does give off a poison that can irritate your eyes or make you ill, toads do not produce warts.

Warts—hard, rough growths on the surface of the skin—are actually caused by certain *viruses*. These viruses live in cells on the surface of the skin and do not invade the tissue underneath.

Some warts disappear as mysteriously as they appeared. Perhaps an immunity to the virus develops. An immunity is a resistance to a disease-causing organism or a harmful substance. If a wart does not go away by itself, medical attention should be sought. Under no circumstances should you try to remove a wart without medical help.

So feel free to handle toads all you want. Their wart-producing reputation is simply nonsense!

C ■ 83

C ■ 83

Laboratory Investigation

DESIGNING AN AQUATIC ENVIRONMENT

BEFORE THE LAB

1. Gather all materials at least one day prior to the investigation. Be sure to have enough guppies and snails for each aquarium. There should be one guppy and one snail for every four liters of water.

2. Identify the classroom locations best suited to setting up an aquarium. These locations should have easy access to a sink. A rubber hose or a bucket will be needed to add water to the aquariums.

3. If you are using an electric aquarium light or electric filter mechanism, make sure the aquarium location is accessible to an electrical outlet.

PRE-LAB DISCUSSION

Have students read the complete laboratory procedure.

• **What is the purpose of the laboratory activity?** (To determine the type of environment that is best suited to guppies.)

• **What type of water conditions must be present for a fish to survive?** (Water type, such as salt water or fresh water, and water temperature are two such conditions.)

• **How should live animals such as guppies be handled?** (Live animals should always be handled with care.)

• **What are the potential dangers of using electricity near water?** (Electric shock can occur. Electric shock can cause serious damage or be fatal.)

Laboratory Investigation

Designing an Aquatic Environment

Problem

What type of environment is best for guppies?

Materials *(per group)*

rectangular aquarium (15 to 20 liters)
aquarium light (optional)
gravel   guppies
metric ruler   aquarium cover
water plants   guppy food
aquarium filter   thermometer
snails   dip net

Procedure 🔺 ⚗ 🐌

1. Wash the aquarium with lukewarm water and place it on a flat surface in indirect sunlight. Do not use soap when washing the aquarium.

2. Rinse the gravel and use it to cover the bottom of the aquarium to a depth of about 3.5 cm.

3. Fill the aquarium about two-thirds full with tap water.

4. Gently place water plants into the aquarium by pushing their roots into the gravel. If you have a filter, place it in the aquarium and turn it on.

5. Add more water until the water level is about 5 cm from the top of the aquarium. Let the aquarium stand for 2 days.

6. Add the snails and guppies to the aquarium. Use one guppy and one snail for every 4 liters of water.

7. Place the cover on top of the aquarium.

8. Keep the temperature of the aquarium between 23°C and 27°C. Feed the guppies a small amount of food each day. Add tap water that has been left standing for 24 hours to the aquarium as needed. Remove any dead plants or animals.

9. Observe the aquarium every day for 2 weeks. Record your observations.

Observations

1. Do the guppies swim alone or in a school?

2. What do you see when you observe the gills of guppies?

3. Describe the reaction of guppies when food is placed in the aquarium.

4. Describe the method snails use to obtain their food.

5. Was there any growth in the water plants? How do you know?

Analysis and Conclusions

1. To what phylum of animals do snails belong? To what phylum of animals do guppies belong? How do you know?

2. How do fishes obtain oxygen?

3. What is the function of the water plants in the aquarium? The function of snails?

4. Why is it important that you do not overfeed the guppies?

5. Why did you allow the tap water to stand for 24 hours?

6. **On Your Own**   Design an experiment to determine how the aquarium would be affected by the following conditions: placing the aquarium in direct sunlight and in darkness and adding guppies.

TEACHING STRATEGIES

1. Point out that some observations may be completed during the course of the investigation, not at any one moment. For example, at a given instant the guppies may not be swimming in schools, yet at most other times they are. It would be incorrect to observe and conclude in this situation that guppies do not swim in schools.

2. Remind students to use caution when handling live animals and fragile equipment. Also point out that they should periodically check the location of any electrical cords throughout the course of the investigation.

DISCOVERY STRATEGIES

Discuss how the investigation relates to the chapter by asking open questions similar to the following:

• **If you wished to increase or decrease the physical activity of the guppies in this investigation, what are some things you might consider doing?** (Answers will vary. Students might suggest increasing or decreasing the temperature of the water because guppies are coldblooded organisms—analyzing, relating concepts.)

Summarizing Key Concepts

3–1 What Is a Vertebrate?

▲ A vertebrate is an animal that has a backbone, or vertebral column. The vertebral column is part of a vertebrate's endoskeleton, or internal skeleton.

▲ All vertebrates belong to the phylum Chordata. At some time during their lives, all chordates have three important characteristics: a nerve cord, a notochord, and a throat with gill slits.

▲ Gills are feathery structures in which the exchange of the gases oxygen and carbon dioxide occurs. Fishes have gills.

▲ Coldblooded animals do not produce much heat. Thus they must rely on their environment for the heat they need.

▲ Warmblooded animals maintain their body temperatures internally as a result of the chemical reactions that occur within their cells.

3–2 Fishes

▲ Fishes are water-dwelling vertebrates that are characterized by scales, fins, and throats with gill slits.

▲ Fertilization in fishes may be external or internal. External fertilization is the process in which a sperm joins with an egg outside the body. Internal fertilization is the process in which a sperm joins with an egg inside the body.

▲ Fishes are placed into three main groups: jawless fishes, cartilaginous fishes, and bony fishes.

▲ Jawless fishes are eellike fishes that lack paired fins, scales, and a backbone.

▲ Cartilaginous fishes have a skeleton of flexible cartilage.

▲ Bony fishes have skeletons of bone. Most have swim bladders, which are gas-filled sacs that give bony fishes their buoyancy.

3–3 Amphibians

▲ Amphibians are vertebrates that are fishlike and that breathe through gills when immature. They live on land and breathe through lungs and moist skin as adults. Their skin also contains many glands, and their bodies lack scales and claws.

▲ Young amphibians have a single-loop circulatory system. Adult amphibians have a double-loop circulatory system.

▲ Fertilization in amphibians may be external or internal.

▲ Metamorphosis is a series of dramatic changes in body form in an amphibian's life cycle.

▲ As adults, frogs and toads develop lungs and legs.

▲ Salamanders and newts are amphibians with tails.

Reviewing Key Terms

Define each term in a complete sentence.

3–1 What Is a Vertebrate?	3–2 Fishes	3–3 Amphibians
vertebrate	external fertilization	metamorphosis
gill	internal fertilization	
coldblooded	swim bladder	
warmblooded		

C ■ 85

ANALYSIS AND CONCLUSIONS

1. Snails belong to the phylum Mollusca because they have soft bodies covered with a hard shell. Fishes belong to the phylum Chordata because they have a backbone and an internal skeleton.

2. Through their gills.

3. Water plants provide oxygen and food for the snails and fishes. Snails help to keep the aquarium glass clean by eating algae, and they also give off carbon dioxide, which is needed by plants during the process of photosynthesis.

4. Excess food would foul the water and kill the fishes.

5. To bring the water to room temperature, to dissolve air in the water, and to let various sediments settle.

6. Experiments will vary, but each should be designed to test the effects of sunlight, darkness, and more guppies.

GOING FURTHER: ENRICHMENT
Part 1

To continue the study of fishes and their environment, have students investigate fishes that require a saltwater environment. Set up another aquarium following the same procedure used previously. In step 3 of the previous procedure, however, fill the aquarium two-thirds full with ocean (salt) water instead of tap water. Various types of tropical fishes could be added to this aquarium. Have students research the temperature range needed by the tropical fishes.

Chapter Review

ALTERNATIVE ASSESSMENT

The *Prentice Hall Science* program includes a variety of testing components and methodologies. Aside from the Chapter Review questions, you may opt to use the Chapter Test or the Computer Test Bank Test in your *Test Book* for assessment of important facts and concepts. In addition, Performance-Based Tests are included in your *Test Book*. These Performance-Based Tests are designed to test science process skills, rather than factual content recall. Since they are not content dependent, Performance-Based Tests can be distributed after students complete a chapter or after they complete the entire textbook.

• **Because guppies are relatively small organisms, what assumption(s) might you make about their birth rate in a natural outdoor environment?** (Students should infer that female guppies would give birth to many guppies at one time or that the number of times female guppies give birth would be relatively frequent—inferring, predicting.)

• **Why?** (Reasons will vary. Students might infer that guppies provide easy and numerous targets for larger predators—relating, generalizing.)

OBSERVATIONS

1. Guppies swim in loosely grouped schools.

2. The gill covers seem to open and close.

3. The guppies swim toward the food.

4. Snails crawl on plants and eat the leaves, as well as algae that may build up along the glass walls of the aquarium.

5. Yes. The aquarium became crowded with plants.

CONTENT REVIEW

Multiple Choice

1. c
2. b
3. a
4. c
5. b
6. d
7. b
8. b
9. c
10. d

True or False

1. T
2. T
3. F, jawless
4. T
5. F, cartilaginous
6. T
7. T
8. F, amphibians

Concept Mapping:

Row 1: Vertebrates
Row 2: Coldblooded
Row 3: Fishes, Amphibians, Birds

CONCEPT MASTERY

1. Fishes have fins to help propel them through the water, swim bladders or air bladders to help them float at different depths, and gills to allow them to take in oxygen from the water.

2. Amphibians lay their eggs in water and, as such, must live near water most of their lives. Many amphibians have a moist skin and need water to help keep their skin moist during dry times of the year.

3. At some time during their lives, chordates have a nerve cord, a notochord, and a throat with gill slits.

4. Adult amphibians have lungs and legs that successfully allow them to live on land.

5. In a single-loop system, blood travels from the heart to the gills to the rest of the body and back to the heart. In a double-loop system, the first loop carries oxygen-poor blood from the lungs back to the heart, and the second loop transports oxygen-rich blood from the heart to the rest of the body and oxygen-poor blood from the body back the the heart.

Content Review

Multiple Choice

Choose the letter of the answer that best completes each statement.

1. All vertebrates have
   a. bony skeletons.
   b. scales.
   c. vertebral columns.
   d. exoskeletons.
2. Which is not a vertebrate?
   a. snake
   c. shark
   b. earthworm
   d. lizard
3. Which group of vertebrates are warm-blooded?
   a. mammals
   c. amphibians
   b. fishes
   d. reptiles
4. Which is best suited for life in water?
   a. toad
   c. trout
   b. newt
   d. frog
5. Which fish lacks jaws, scales, and paired fins?
   a. paddlefish
   c. shark
   b. hagfish
   d. skate

6. Which is a cartilaginous fish?
   a. electric eel
   c. trout
   b. lungfish
   d. ray
7. Bony fishes can float at different levels in the water because they have
   a. backbones.
   c. fins.
   b. swim bladders.
   d. gills.
8. Amphibians must lay their eggs
   a. on land.
   c. in nests.
   b. in water.
   d. in shells.
9. Immature frogs breathe through their
   a. lungs.
   c. gills.
   b. skin.
   d. mouth.
10. Which is not an amphibian?
    a. frog
    c. newt
    b. toad
    d. lamprey

True or False

If the statement is true, write "true." If it is false, change the underlined word or words to make the statement true.

1. <u>Vertebrates</u> are members of the phylum Chordata.
2. All vertebrates have an <u>endoskeleton</u>.
3. The <u>cartilaginous</u> fishes are the most primitive group of fishes.
4. The lamprey is a <u>jawless</u> fish.
5. Sharks are <u>bony</u> fishes.
6. The skin of a toad is <u>drier</u> than that of a frog.
7. Adult amphibians obtain most of their oxygen through their <u>lungs</u>.
8. Newts and salamanders are <u>fishes</u> with tails.

Concept Mapping

Complete the following concept map for Section 3–1. Refer to pages C6–C7 to construct a concept map for the entire chapter.

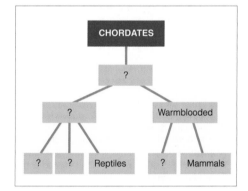

6. As a fish moves, it can sense changes in the movement of water, enabling it to detect prey or avoid objects in its path.

7. In many fishes with external fertilization, the eggs are abandoned after they are fertilized. In fishes with internal fertilization, the unborn offspring are carried in the mouth of one of the parents during the hatching period.

8. Jawless fishes are the most primitive fishes, lacking jaws, scales, bones, and paired fins. Cartilaginous fishes have skeletons made of cartilage and scales. Bony fishes have skeletons made of bones, paired fins, and swim bladders.

9. The survival rate for internally fertilized vertebrates is significantly higher than the rate for externally fertilized vertebrates. To compensate and help ensure survival, externally fertilized vertebrates usually produce greater numbers of eggs.

Concept Mastery

Discuss each of the following in a brief paragraph.

1. **What** adaptations have fishes developed that enable them to live in water?
2. Explain why amphibians must live in a moist environment.
3. Describe the main characteristics of chordates.
4. What adaptations have adult amphibians developed that allow them to live successfully on land?
5. How does a double-loop circulatory system differ from a single-loop circulatory system?

6. What is the "distant touch" system in fishes?
7. How do fishes care for their young?
8. Compare the three groups of fishes.
9. Hypothesize why those vertebrates that reproduce by internal fertilization tend to produce fewer eggs than do those animals that reproduce by external fertilization.
10. Describe metamorphosis in frogs.
11. Explain what happens to frogs and toads when they go into hibernation.

Critical Thinking and Problem Solving

Use the skills you have developed in this chapter to complete each of the following.

1. **Relating facts** Why would you never find frogs living in Antarctica?
2. **Making inferences** People who fish often use a variety of artificial lures. Explain how these lures could attract fishes.
3. **Developing a hypothesis** Some fishes have light colors on their bottom surfaces and dark colors on their top surfaces. Develop a hypothesis to explain how this coloration could be an advantage.
4. **Relating facts** When a raccoon catches a toad, it usually wipes the amphibian along the ground before eating it. Suggest a reason for this strange behavior.
5. **Applying concepts** A female bullfrog can produce as many as 25,000 eggs in a year. Explain why the Earth is not overrun with bullfrogs.
6. **Designing an experiment** Design an experiment in which you determine whether salamanders are able to detect sound. Be sure to include a variable and a control in your experiment.

7. **Using the writing process** Pretend that you received a letter from a friend who lives in another state. She has told you about a small four-legged cold-blooded vertebrate that she found. She wants to know if it is a frog or a salamander. Write a letter telling her how she can determine what her animal is.

---

10. Frog eggs are fertilized externally, developing in water and hatching into young with tails (tadpoles). Gradually, the tadpoles grow limbs and begin to lose their tails as they develop into adult frogs.

11. During hibernation, the body processes of frogs and toads slow down so that each animal can live on food stored in its body and breathe through its skin.

CRITICAL THINKING AND PROBLEM SOLVING

1. Frogs are coldblooded and cannot adapt to the cold environment of the Antarctic. Their body temperature would fall too low, despite any behavioral attempts to keep warm such as basking in the sun.

2. People who fish use lures that simulate the colors and movements of prey that fishes normally hunt for food. As such, fishes cannot distinguish the lure from the real food.

3. The best defense mechanism many fishes have is to not be seen. Another fish looking down on the dark coloring of the top of a fish may not notice the fish because water appears darker when looking down. When looking up, the water appears lighter, and a fish with a light-colored bottom would be less obvious.

4. The toad may secrete a poisonous liquid that might bother the raccoon. By wiping the toad on the ground, the raccoon removes this liquid before eating the toad.

5. The bullfrog, like many organisms, produces more eggs than might possibly hatch. Many eggs are destroyed or eaten. By laying so many eggs, the bullfrog ensures that at least some of the eggs will hatch and mature.

6. Experiments will vary but should be logical and reflect the scientific method.

7. Letters will vary. Though the methods chosen by students to determine the species of an animal might vary, each method should be scientifically accurate.

KEEPING A PORTFOLIO

You might want to assign some of the Concept Mastery and Critical Thinking and Problem Solving questions as homework and have students include their responses to unassigned questions in their portfolio. Students should be encouraged to include both the question and the answer in their portfolio.

ISSUES IN SCIENCE

The following issues can be used as a springboard for discussion or given as a writing assignment:

The only method that has been successful in reducing the number of lampreys in the Great Lakes is to poison their young in the freshwater streams where they develop. A favorite fish of people who fish is the coho salmon, whose young develop in the same streams. These people believe that the poison may also be responsible for reducing the number of coho. Some have urged that the practice of poisoning lampreys be stopped. What is your opinion?

SECTION	HANDS-ON ACTIVITIES
4–1 Reptiles pages C90–C102 Multicultural Opportunity 4–1, p. C90 ESL Strategy 4–1, p. C90	**Student Edition** ACTIVITY (Discovering): Eggs-amination, p. C92 ACTIVITY (Doing): Snakes, p. C94 **Teacher Edition** New Light on Old Bones, p. C88d **Laboratory Manual** Adaptations of Lizards, p. C87
4–2 Birds pages C103–C113 Multicultural Opportunity 4–2, p. C103 ESL Strategy 4–2, p. C103	**Student Edition** ACTIVITY (Discovering): Comparing Feathers, p. C104 LABORATORY INVESTIGATION: Owl Pellets, p. C114 ACTIVITY BANK: Do Oil and Water Mix? p. C163 ACTIVITY BANK: An Eggs-aggeration, p. C164 ACTIVITY BANK: Strictly for the Birds, p. C165 **Teacher Edition** Making Bones About It, p. C88d **Activity Book** CHAPTER DISCOVERY: Birds of a Feather, p. C73 **Laboratory Manual** Examining Bird Adaptations, p. C93
Chapter Review pages C114–C117	

OUTSIDE TEACHER RESOURCES

Books

Blassingame, Wyatt. *Wonders of Alligators and Crocodiles,* Dodd Mead.

Burnie, David. *Birds,* Knopf.

Fichter, George S. *Poisonous Snakes,* Franklin Watts.

Graham, Ada, and Frank Graham. *Alligators,* Delacorte.

Hendrick, Paula. *Saving America's Birds,* Lothrop.

Roever, J. M. *Snake Secrets,* Walker.

Roop, Peter, and Connie Roop. *Seasons of the Cranes,* Walker.

Ryder, Hope. *America's Bald Eagle,* Putnam.

Scott, Jack Denton. *Swans,* Putnam.

Simon, Hilda. *Snakes: The Facts and Folklore,* Viking.

Yolen, Jane. *Bird Watch,* Philomel.

Yount, Lisa. *Too Hot, Too Cold, Just Right: How Animals Control Their Temperatures,* Walker.

Audiovisuals

Birds in Winter, video, EBE

Life in a Bird's Nest, filmstrip with cassette, SVE

Reptiles, video, EBE

Reptiles, video, National Geographic

Snakes and How They Live, video, AIMS Media

OTHER ACTIVITIES	MEDIA AND TECHNOLOGY
Student Edition ACTIVITY (Reading): It All Depends on Your Point of View, p. C91 **Activity Book** ACTIVITY: Observing Reptile Locomotion, p. C77 ACTIVITY: Unusual Coldblooded Vertebrates, p. C83 ACTIVITY: Fact or Fiction? p. C85 ACTIVITY: Investigating How Reptiles Regulate the Temperature of Their Body, p. C87 **Review and Reinforcement Guide** Section 4–1, p. C49	**Video** Reptiles (Supplemental) **English/Spanish Audiotapes** Section 4–1
Student Edition ACTIVITY (Reading): Rara Aves, p. C109 ACTIVITY (Thinking): A Flock of Phrases, p. C113 **Activity Book** ACTIVITY: What Animal Was Here? p. C81 **Review and Reinforcement Guide** Section 4–2, p. C55	**Interactive Videodisc/CD ROM** Paul ParkRanger and the Mystery of the Disappearing Ducks **Video/Videodisc** The Eagle's Story (Supplemental) Birds (Supplemental) **English/Spanish Audiotapes** Section 4–2
Test Book Chapter Test, p. C69 Performance-Based Tests, p. C109	**Test Book** Computer Test Bank Test, p. C77

*All materials in the Chapter Planning Guide Grid are available as part of the Prentice Hall Science Learning System.

CHAPTER OVERVIEW

Reptiles are vertebrate animals that have adaptations, such as lungs, dry and scaly skin, and amniotic eggs, that enable them to live their entire lives out of water. Some reptile species bear live young; others lay eggs that develop outside the mother's body. All reptiles practice internal fertilization.

Examples of reptiles include lizards, snakes, turtles, alligators, and crocodiles. Reptiles are coldblooded organisms, meaning they obtain heat from outside their bodies. Depending on the particular environment, being coldblooded can be an advantage or disadvantage to the survival strategies of an organism.

Birds are warmblooded, reptilelike animals with an outer covering of feathers, two legs used for walking or perching, and front limbs that have modified into wings. Fossil evidence confirms the evolution of birds from ancient reptiles.

Bird feathers are generally of two different types: contour and down. Birds' wings show many variations that involve flight. Hollow bones and large chest muscles are found in flying birds. Birds practice internal fertilization and lay eggs that are usually incubated until they hatch.

4–1 REPTILES
THEMATIC FOCUS

The purpose of this section is to describe the physical and behavioral characteristics of the different types of reptiles. Reptiles are vertebrate animals that have special adaptations that enable them to live on land. Reptilian skin is completely covered by a tough, dry, relatively thick layer of scales. These scales are formed by the outermost layer of the skin and are made of dead, flattened cells that contain the same hard, tough substance that is found in human fingernails. Reptilian scales form an unbroken waterproof covering that helps to prevent reptiles from water loss and drying out.

The kidneys of reptiles are adapted to prevent excess water loss. The kidneys of reptiles concentrate nitrogen-containing waste products so that as little water as possible is lost when wastes are eliminated.

Another reptilian adaptation involves the respiratory and circulatory systems. In some reptiles such as crocodiles, the circulatory system contains two separate loops. This system does not allow oxygen-poor blood in one loop to mix with oxygen-rich blood in the other. Because the bloods never mix, oxygen is delivered more efficiently to the cells of the body.

Reptiles have another important adaptation for living on land—their eggs. Many reptile eggs have a hard shell, which is similar to that of a chicken egg. Others have shells that are more flexible and leathery. The shell contains pores that are large enough to allow gases to pass, yet small enough to prevent water

from easily escaping. The reptilian egg is of great evolutionary importance because it freed reptiles from their dependence on water for reproduction and development, and it clearly links reptiles to the vertebrates that evolved from them: birds and mammals.

Typical reptiles include lizards and snakes, turtles, and alligators and crocodiles. Because of their adaptation to many different terrestrial environments, reptiles exhibit numerous variations in structure and behavior. All body systems of reptiles, however, are adapted to a successful life on land.

The themes that can be focused on in this section are evolution, patterns of change, scale and structure, unity and diversity, systems and interactions, and stability.

*Evolution: Reptiles appeared hundreds of millions of years ago, soon after the first amphibians. The death of all the great and terrible "lizards"—dinosaurs and their giant contemporaries—left many niches open for mammals, both on the land and in the sea.

*Patterns of change: Reptiles, unlike amphibians, do not undergo metamorphosis. Development is essentially completed while in the egg.

*Scale and structure: The shell and membranes of a reptilian egg make it possible for the egg to be laid on land.

*Unity and diversity: Reptiles may be egg-laying or live-bearing, and the amount of parental care among reptiles varies greatly.

Stability: Certain reptiles return year after year to the same places to breed and/or lay their eggs.

Systems and interactions: The body systems of reptiles make them and their descendants better suited to life on land than amphibians are, enabling them to inhabit different types of land environments.

PERFORMANCE OBJECTIVES 4–1

1. Describe several characteristics of reptiles.

2. Relate the structure and function of reptiles to their success in dry environments.

3. Compare the similarities and differences of some specific types of reptiles.

4–2 BIRDS
THEMATIC FOCUS

The purpose of this section is to describe the physical and behavioral characteristics of birds. Even though the group of birds is a diverse collection of living things, there are several characteristics birds share: They are warmblooded reptilelike animals with an outer covering of feathers, they have two legs used for walking or perching, and they have front limbs modified into wings.

The single most important characteristic that separates birds from reptiles is feathers. Feathers help a bird fly and also keep warm. Birds have several different kinds of feathers. Because many feathers are hollow, they are both light and strong. Contour feathers are the most noticeable feathers on a bird; they provide the lifting force and balance necessary for flight. Down feathers grow underneath and between the contour feathers, trapping air and insulating the bird.

Birds have high metabolic rates and burn many calories just to remain warm. For that reason, birds need to eat large amounts of food.

The respiratory system of birds is extremely efficient at taking in oxygen and eliminating carbon dioxide—the high metabolic rate of birds demands an efficient gas exchange system. The reason for this efficiency is that bird lungs are connected at both anterior and posterior to large air sacs in the body cavity and bones. The large air sacs also serve to make a bird's body more buoyant, allowing the bird to fly more easily.

Despite the derogatory term *birdbrain*, birds are really intelligent animals. Their brains are large and well developed. They also have excellent senses, especially eyesight. They may be able to visually distinguish colors better than humans. Although birds lack external ears, they have ear openings in their head. Most bird species can hear quite well. Many migratory birds can hear the pounding of waves on a shoreline even when they are many kilometers away. Although the exact processes of navigation are not thoroughly understood, some migratory birds use a magnetic sense to navigate. This magnetic sense, located somewhere in the head, operates like a built-in compass, responding to the Earth's magnetic field. Many migratory birds use a combination of keen eyesight, instinct, and a built-in clock to navigate by the sun and stars.

The reproductive system of birds is similar to that of reptiles. Internal fertilization produces fertilized eggs that have hard outside shells, with an internal structure and membranes similar to those found in reptile eggs. Unlike many reptiles, most birds incubate their eggs until the eggs hatch. Birds also have fascinating courtship and mating behaviors. The males of some species—peacocks,

for example—use brightly colored feathers to attract females and warn off other males during the breeding season. Male weaverbirds construct a nest that is examined by a prospective mate for soundness and craftsmanship. A male penguin does not construct a large or elaborate structure; instead, he presents his mate with a pebble, indicating he is ready to breed and care for young.

The themes that can be focused on in this section are evolution, patterns of change, scale and structure, unity and diversity, systems and interactions, stability, and energy.

*Evolution: Birds evolved comparatively recently from reptile ancestors, which were probably dinosaurs.

*Patterns of change: Birds migrate in response to seasonal changes in food supply.

*Scale and structure: The shapes and colors of feathers help them perform their functions.

*Unity and diversity: Although they vary substantially, all vertebrates with feathers are classified as birds.

Stability: Certain birds return year after year to the same places to breed and/or lay their eggs. Feathers help to insulate birds.

Systems and interactions: Bird body systems show many adaptations for flight. Birds communicate with other members of their species and of their community.

Energy: In order to meet the energy demands of warmbloodedness and flight, birds must acquire significantly large quantities of energy in the form of food.

PERFORMANCE OBJECTIVES 4–2

1. Identify the major characteristics of birds.
2. Describe ways in which the form of birds shows adaptations for flight.
3. Describe the four main categories of birds.
4. List several ways in which birds might navigate during migration.

SCIENCE TERMS 4–2

feather p. C103
contour feather p. C104
down p. C104
territory p. C108
migrate p. C109

——— Discovery *Learning* ———

TEACHER DEMONSTRATIONS MODELING

New Light on Old Bones

Though many species of living things have become extinct throughout history, students are probably most familiar with the extinction of reptile dinosaurs. Gather all the pictures you can find of dinosaurs and distribute or display the pictures throughout the classroom.

Have a group of interested students conduct a formal debate on the different hypotheses concerning the extinction of dinosaurs. Have students who are not participating in the debate decide which hypothesis is most feasible.

Making Bones About It

Before the demonstration, collect and clean some leftover bones from a cooked chicken. Saw the bones in half. Collect and clean bones from a different animal, such as beef or pork bones. Saw the bones in half. Boil all bones to remove excess fat and muscle tissue.

Distribute the bones to students.
• **What is the difference between the two kinds of bones?** (Accept all logical answers. Students might suggest that the bones are of different sizes and shapes.)
• **Which bone do you think came from a bird? Why do you think so?** (Accept all logical answers. Lead students to suggest that the hollow bones weigh less and should be better for flying.)
• **Which bone do you think came from a large reptile or mammal? Why do you think so?** (Accept all logical answers. Lead students to suggest that the solid bones are stronger and would support a heavier body.)

CHAPTER 4
Reptiles and Birds

INTEGRATING SCIENCE

This life science chapter provides you with numerous opportunities to integrate other areas of science, as well as other disciplines, into your curriculum. Blue numbered annotations on the student page and integration notes on the teacher wraparound pages alert you to areas of possible integration.

In this chapter you can integrate life science and classification (pp. 90, 110), life science and evolution (pp. 91, 92, 97, 103, 106, 111), language arts (pp. 91, 103, 104, 109), life science and cells (p. 94), social studies (p. 95), physical education (p. 103), life science and ecology (p. 108), and dance (p. 113).

SCIENCE, TECHNOLOGY, AND SOCIETY/COOPERATIVE LEARNING

Historically, the animals that people shared their environment with provided people with food, clothing, and sometimes shelter. People used skins from various animals for clothing and shelter, bones for tools, and meat for food. Today, animals still provide people with food and clothing, but some animals are hunted and killed only for their exotic skins.

Many species of reptiles and a few species of birds are valued for their hides and/or their feathers. Alligators, crocodiles, lizards, snakes, emus, and ostriches are a few of many species used in the manufacturing of expensive belts, handbags, luggage, shoes and boots, feather "boas," or feather dusters. Throughout the swamps of the southern United States, alligators were hunted and killed for the hundreds of dollars that their hides would be worth. Hunted almost to the point of extinction, the American alligator came under the protection of the federal government. The government limited the amount of hunting that could be done, carefully controlled the licensing of hunters, and seized products made from hides obtained by poachers, or illegal hunters. The protection of the alligator in the United States has been so successful that the alligator population is

INTRODUCING CHAPTER 4

DISCOVERY LEARNING

▶ *Activity Book*
Begin teaching the chapter by using the Chapter 4 Discovery Activity from the *Activity Book*. Using this activity, students will explore some of the characteristics of bird feathers.

USING THE TEXTBOOK

Have students observe the photograph on page C88.
• **Describe what you see.** (A dinosaur scene.)
• **What can you infer about the climate of the region shown?** (The climate appears to be tropical.)
• **Does the area appear to be sparsely or densely populated with living things?** (The area appears densely populated.)
• **How do the sizes of the animals seem**

Reptiles and Birds

Guide for Reading

After you read the following sections, you will be able to

4–1 Reptiles

- Describe the adaptations that allow reptiles to live their entire lives on land.
- Explain how reptiles carry out their major life functions.

4–2 Birds

- Describe the characteristics of birds.
- Discuss the ways in which birds perform their major life functions.

Imagine that you have just journeyed back in time. You step out of your time machine and enter the world of 150 million years ago.

As you look around at this strange world of the past, it becomes clear to you why this part of Earth's history is sometimes called the Age of Reptiles: Reptiles are the dominant form of life. Long-beaked reptiles soar on narrow wings in the sky above you. Porpoiselike reptiles come to the surface of the ocean for a breath of air, then dive back into the depths with a flick of their fishlike tail. On land, the reptiles known as dinosaurs roam among forests of tree ferns, conifers, and cycads. A rhinoceros-sized dinosaur with huge pointed plates on its back swishes its spike-tipped tail as you approach. Fierce meat-eating dinosaurs as tall as giraffes run swiftly on their two hind legs, pursuing a herd of plant-eating dinosaurs that are, astoundingly, even taller!

Although most of the reptiles of 150 million years ago have died out, some types have survived to the present. What are reptiles? What sorts of reptiles are alive today? Where do reptiles fit in the evolutionary tree of life? Read on and learn about reptiles and birds—living relatives of dinosaurs.

Journal *Activity*

You and Your World In your journal, list ten of the most important facts that you already know about reptiles. Make a similar list for birds. Compare your list with a friend's list and discuss any differences. After you have finished studying this chapter, look over your list. Make any changes you like and briefly note why the changes were made.

About 150 million years ago, reptiles dominated the Earth.

---

to compare to the surrounding landscape? (The animals appear enormous when compared to the surrounding landscape.)

• **What characteristics do living things of this period have in common?** (Accept reasonable responses.)

• **What characteristics make these animals well adapted to their lifestyles?** (Answers will vary. Students might suggest that the long necks of plant-eaters enhance their ability to obtain food.)

• **How do scientists know what dinosaurs looked like, given that dinosaurs no longer exist on Earth?** (Accept all logical answers. Students might cite fossil evidence.)

• **What are some modern reptiles that resemble ancient dinosaurs?** (Answers may include lizards, crocodiles, alligators, and turtles.)

Point out that land-dwelling reptiles entered a period of rapid evolution as the Earth's climate became drier. Stress that like dinosaurs modern reptiles and birds evolved from ancient reptiles.

---

once again thriving, and the government has been able to establish regular hunting seasons in which a limited number of alligators can be killed. Other species of reptiles and birds have not been as fortunate as the American alligator.

Today, people can purchase alligator-like leather that is really cowhide, processed to look like alligator or many other of the exotic skins used in fashion accessories and shoes. Modern technology has also created a number of synthetic materials that resemble the skins of reptiles and the feathers of birds. Some people wonder why certain species of living animals continue to be hunted for their skins.

Cooperative learning: Using preassigned groups or randomly selected teams, have groups complete one of the following assignments.

• Randomly assign groups one of the reptiles or birds mentioned in this activity or other animals that are valued for their body parts, which are used to meet people's need for luxury or novelty items. Each group should identify the way(s) in which the animal is "valuable." You might encourage groups to consider the organism's ecological, economic, medicinal, aesthetic value, and so on, in their discussions.

• Many people have an unreasonable fear of snakes and have the attitude that the "only good snake is a dead snake." Acting as the public relations committee for the imaginary group People for the Reptile Way, create an advertising campaign designed to educate people about snakes and the service that they perform for our society.

See Cooperative Learning in the *Teacher's Desk Reference.*

JOURNAL ACTIVITY

You may want to use the Journal Activity as the basis of a class discussion. After the discussion, you might choose to extend the activity by having students use the chalkboard to create a master list of ten different reptiles and ten different birds. Then have volunteers describe how the environment and they personally would be affected if these 20 organisms suddenly became extinct. Students should be instructed to keep their Journal Activity in their portfolio.

4-1 Reptiles

The tortoise plays a role in the folklore of many cultures. Proverbs and sayings as well as legends often refer to the tortoise. Students may be familiar with the story "The Tortoise and the Hare," for example, which tells how a slow but steady tortoise won a race against a fast but lazy hare. This ancient tale, retold by Aesop and later by writers of fables throughout the world, speaks about determination and hard work as they are reflected in the value system of various cultures. Students may want to tell the class about other animals used as symbols for values that are important in their ethnic cultures.

ESL STRATEGY 4-1

Give students the following horizontal chart headings: Physical Features, Egg-Laying Process, Lifestyle.

Then give students the following vertical chart headings: Amphibians, Reptiles.

Have students create a chart that uses the given headings to compare reptiles and amphibians. Students should complete their charts by choosing from the following information: undergo metamorphosis; live on land or water; thin, moist skin; double-loop circulatory system; live on land; coldblooded; internal fertilization; scaly skin; lungs; eggs require damp environment; internal/external fertilization; eggs can be laid anywhere; better developed lungs; do not undergo metamorphosis; young complete development inside eggshell; kidneys prevent water loss; most lay eggs, but some bear live young; live in water when young.

Some of the information given may apply to both amphibians and reptiles.

Figure 4-1 *The four living scientific groups of reptiles are represented by the spectacled caiman (top left), Florida red-bellied turtle (top right), tuatara (bottom left), and marine iguana (bottom right).* ❶

4-1 Reptiles

On the barren, windswept shoreline of the Galapagos Islands in the Pacific Ocean, a group of large lizards called marine iguanas (ih-GWAH-nuhz) cling to the rocks. Wave after wave splash against the rocks, but the iguanas do not let go. Suddenly, the iguanas plunge into the cold sea. Their lashing tails and webbed feet propel them through the water as they dive for the seaweed on which they feed. With their bodies chilled by the water, the iguanas soon scramble back onto the rocks to warm up.

Iguanas are just one example of the group of vertebrates known as reptiles. Other reptiles include snakes, turtles, crocodiles, and extinct (no longer living) animals such as dinosaurs and pterodactyls.

Reptiles are vertebrates that have lungs, scaly skin, and a special type of egg. These characteristics, which you will soon read about in more detail, make it possible for reptiles to spend their entire lives out of water. Another important characteristic

---

TEACHING STRATEGY 4-1

FOCUS/MOTIVATION

People are sometimes naturally fearful of reptiles. Snakes, in particular, have a bad reputation and are the subject of much dislike and many misunderstandings.

• **What have you heard about snakes?** (Accept all answers.)

As students make their statements, discuss whether each statement is true or false. All false statements should be refuted.

The characteristics of a reptile can be introduced by obtaining several live specimens, such as a turtle, small lizard, horned toad, or small nonpoisonous snake. Allow students to observe and gently handle the specimens to become comfortable with them.

• **What characteristics do these animals have in common?** (Accept all logical an-

of living reptiles involves the way they control their body temperature. **All living reptiles are cold-blooded.** Do you recall from Chapter 3 how cold-blooded animals control their body temperature? ②

Reptiles appeared hundreds of millions of years ago, soon after the first amphibians. As you can see in Figure 4–2, the first reptiles were large, fat, short-legged animals that resembled a cross between a lizard and a toad. Although the first reptiles looked a lot like the ancient amphibians that dominated the Earth at that time, there were several important differences. These differences enabled early reptiles and their descendants (modern reptiles, birds, and mammals) to inhabit all sorts of land environments. As you read about the characteristics of modern reptiles, focus on the ways in which these characteristics make reptiles better suited to life on land than amphibians.

Reptiles have skins that are completely covered by a tough, dry, relatively thick layer of scales. These scales are formed by the outermost layer of the skin. They are made of dead, flattened cells that contain the same hard, tough substance found in your fingernails. The scales form an unbroken waterproof covering that helps to prevent drying out.

Although the waterproof skin helps to prevent excess water loss, it also makes it impossible for a typical reptile to breathe through its skin. A thin, moist membrane is needed to transport oxygen from the environment into the body and carbon dioxide from the body to the environment. Because a typical reptile (unlike a typical amphibian) does not have skin that is thin and moist, it depends entirely on its lungs for gas exchange. Would you expect the lungs of reptiles to be more complex or less complex than those of amphibians? Why? ③

The waterproof scaly skin is not the only reptilian adaptation for preventing excess water loss. The kidneys of reptiles concentrate nitrogen-containing waste products so that as little water as possible is lost when wastes are eliminated.

ACTIVITY READING

It All Depends on Your Point of View

We humans tend to believe that being warm-blooded, live-bearing, and mammalian are the best things to be. But suppose that an intelligent cold-blooded reptilian life form had evolved. Would it have viewed things the same way? ③

For one writer's view of a reptile-dominated Earth, read the novel *West of Eden* by Harry Harrison.

swers. Students might suggest that they are vertebrates with lungs.)

• **What characteristics make these animals well adapted to life on land?** (Accept all logical answers. Students might suggest lungs and a protective skin.)

CONTENT DEVELOPMENT

Have students identify the characteristics of reptiles. As students describe each characteristic, explain how the characteristic allows the reptile to be more successful as a land dweller.

Remind students that reptiles are vertebrates that have lungs, scaly skin, and a special type of egg. Also, these vertebrates share the trait of coldbloodedness with fish and amphibians.

● ● ● ● **Integration** ● ● ● ●

Use Figure 4–1 to integrate classification concepts into your lesson.

Use the discussion about the first reptiles on Earth to integrate concepts of evolution into your lesson.

REINFORCEMENT/RETEACHING

Have students review the characteristics of amphibians and the eggs they produce.

ACTIVITY
DISCOVERING
EGGS-AMINATION

Discovery Learning

Skills: Making observations, relating concepts

Materials: magnifying glass, egg, bowl, water

In this activity students should discover that the inside of the blunt end of the egg contains a thin membrane covering what appears to be a pocket of air. When the eggshell containing the membrane is filled with water, the membrane does not collapse; it moves to a slightly different location within the shell, keeping the "air bubble" intact. The different parts of the egg, such as the yolk, albumen (or egg white), and the shell all combine to feed and protect the growing embryo. A description of how the structure of an egg and its shell help an egg perform its functions will vary from student to student. Students might suggest that the structure of an egg contains all the nutrients that a growing embryo will need and a covering or shell that protects the embryo and prevents the contents of the egg from drying out.

4-1 (continued)

CONTENT DEVELOPMENT

Explain that an important reason for the success of reptiles on land was their development of a more efficient respiratory system. Have students recall that most amphibians take in oxygen and release waste gases primarily through their moist skin. This method of respiration works only as long as the amphibian's skin remains damp. In dry land environments, breathing through moist skin is not an option. The dry skin of reptiles, which prevents water loss, also prevents gases from moving through. Stress that to exchange gases with the environment, reptiles usually have two efficient lungs. This respiratory adaptation contributes to the success of reptiles on land.

Figure 4–3 *As new skin grows in, land vertebrates shed bits of the old, dead, outermost layers of skin. Reptiles such as the chameleon shed their old skin all at once (left). The outermost layer of skin forms a chameleon's horns (center), a rattlesnake's rattle (right), and other structures.*

ACTIVITY
DISCOVERING

Eggs-amination

1. Obtain a fresh chicken egg. Examine it with a magnifying glass.
2. Gently crack the egg into a bowl.
3. Examine the inside of the shell. Look at the blunt end of the egg. What do you find there? Fill one half of the eggshell with water. What do you observe?
4. Examine the contents of the egg. What are the functions of the different parts? What part is the egg cell?
5. Look for a small white spot on the yolk. This marks the spot where the embryo would have developed if the egg had been fertilized.

■ How does the structure of the egg and its shell help it to perform its functions?

Reptiles, like adult amphibians, have a double-loop circulatory system. In some reptiles—crocodiles and their relatives, to be exact—the two loops are completely separate. The oxygen-poor blood in one loop never mixes with the oxygen-rich blood in the other loop. Because the bloods never mix, oxygen is delivered more efficiently to the cells of the body. How does this double-loop system differ from the double-loop system in amphibians? ❶

Reptiles have a brain and nervous system quite similar to that of an amphibian, although the brain may be slightly better developed. Most reptile sense organs are well developed, although there are some exceptions. For example, snakes are deaf, and certain burrowing lizards lack eyes.

The scaly skin, improved breathing system, and water-conserving method of waste elimination helped to make early reptiles better suited for life on land than the amphibians that had come before them. But perhaps the reptiles' most important adaptation for living on land was their special egg.

The eggs of fishes and amphibians are delicate sacs that contain stored food and a developing organism. These eggs dry out easily and thus require a watery or extremely damp environment in which to develop. In contrast, the eggs of reptiles can be laid under forest logs, in beach sand, or in cracks in desert rocks. Why is this so? ❷

● ● ● ● **Integration** ● ● ● ●

Use the discussion of reptile adaptations to integrate concepts of evolution into your lesson.

CONTENT DEVELOPMENT

Explain that a major advance of reptiles over amphibians is the type of egg that reptiles are able to produce. Point out that unlike amphibian eggs, which almost always need to develop in water, reptilian eggs are surrounded by a shell

and several membranes that together create a protected environment in which the embryo can develop. (Reptilian eggs are named *amniotic eggs* for one of those membranes.) In addition to a shell and membranes, amniotic eggs also contain a substantial yolk. The yolk is rich in nutrients that the developing embryo uses until it is ready to hatch.

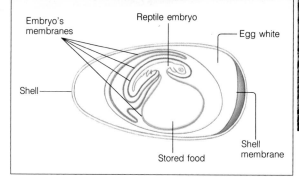

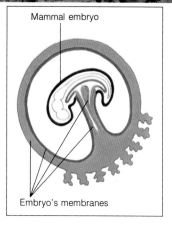

Mammal embryo

Embryo's membranes

Figure 4–4 *These hognose snakes seem to be pleased by their first glimpse of the world. The membranes that surround reptiles as they develop within the egg are shown in the diagram on the left. These membranes also surround developing mammals, as shown in the diagram on the right.*

Figure 4–4 shows the structures that make it possible for the embryos in reptile eggs to develop on dry land. As you can see, the egg is surrounded by a protective shell that prevents the contents of the egg from drying out. Most reptile eggs have a shell that is tough but a bit flexible—sort of like leather. A few reptile eggs have a hard shell similar to the one on a chicken egg. Although the shell looks solid, it is actually dotted with tiny holes large enough for gases to move in and out but small enough to prevent water from easily escaping. Within the shell are several membranes and a watery fluid, which along with the shell provide further protection for the developing embryo.

The reptilian egg is of great evolutionary importance for two reasons. First, it freed vertebrates from their dependence on water for reproduction and development. Second, it clearly links reptiles to the vertebrates that evolved from them: birds and mammals. Bird eggs have the same basic structure as reptile eggs. And the same membranes that protect and support a reptile or a bird embryo also protect and support mammal embryos.

Fertilization in reptiles is internal. Recall that internal fertilization means that a sperm cell joins with an egg cell inside the parent's body. Why is it necessary for animals with tough, waterproof shells on their eggs to have internal fertilization? ❸

C ■ 93

FACTS AND FIGURES

THE OLDEST FOSSIL REPTILE

In late 1988, the oldest-known fossil reptile was discovered in Scotland. This fossil was determined to be 340 million years old—40 million years older than what was previously thought to be the oldest reptile.

• **Could a reptilian egg survive if it were laid on land?** (Yes.)
• **Why?** (The covering of the egg, or shell, prevents the egg from drying out.)

ENRICHMENT

Students may not connect the ability to fly with reptiles. Scientists, however, have found fossil evidence of flying animals that lived during the Mesozoic Era. Pterodactyls were flying reptiles. They had wings made of thin pieces of skin attached to their front legs and body, and they had a long, thin beak with sharp, curved teeth.

GUIDED PRACTICE

Skills Development

Skill: Making comparisons

To reinforce the concept of reptilian eggs, ask:
• **In what ways are the eggs of reptiles different from those of amphibians?** (Reptile eggs have a leathery, protective shell, amphibian eggs do not.)
• **Why is this covering so important?** (The covering prevents the eggs from drying out.)

• **Where are reptilian eggs fertilized?** (The eggs of reptiles are fertilized inside the mother's body.)
• **How are amphibian eggs fertilized?** (The female usually releases her eggs into a warm, moist environment. The male then deposits his sperm on top of the eggs.)
• **Could an amphibian egg survive if it were laid on land?** (No.)
• **Why not?** (The egg would dry out on land.)

ACTIVITY
DOING

SNAKES

Skills: Relating facts, applying concepts

The pet store owner should advise students about the safest and easiest kinds of snakes to keep as pets. **CAUTION:** *No one should keep as a pet any snake that was caught in the wild.* Most of the snakes in a pet shop are tame and are used to being handled. Students might discover and relate in their reports that snake cages should be closed securely because most snakes are adept at escaping through even the smallest opening. Snakes should receive proper ventilation and also be placed in warm places that are free of drafts. A bowl of water should be placed in the cage, making sure that the bowl is heavy enough so that the snake cannot tip it over. Most pet snakes eat live food such as rodents. This may upset some people, especially young children, so these people should not be present at feeding time. Because snakes eat infrequently, it is relatively easy to keep their cages clean. A layer of crushed corn cob, sold in pet stores, makes a good "bedding" for the cage.

4–1 (continued)

CONTENT DEVELOPMENT

Explain that newly hatched reptiles resemble small adults. Virtually all reptiles reproduce through internal fertilization, which means that a male reptile deposits sperm into the body of the female. It is extremely difficult to determine the sex of a reptile.

● ● ● ● **Integration** ● ● ● ●

Use the description of a reptile egg to integrate concepts of cells into your lesson.

ENRICHMENT

Once fertilization occurs, different reptile species treat their eggs in very different ways. Many species of reptiles are oviparous, laying eggs that develop outside the mother's body. Some species bury their eggs, leaving the hatched

Figure 4–5 *Young reptiles, such as the baby box turtle and newly born copperheads, look like miniature versions of their parents. How does this pattern of development differ from that in most amphibians?* ❶

ACTIVITY

Snakes

Visit a pet store that sells snakes. Observe the various types of snakes that are for sale. Interview the pet shop owner or one of the workers. Find out the feeding habits of each snake and how to care for one at home. Present your findings to the class in an oral report accompanied by a poster showing your observations.

young without parental care. Some species provide minimal care by incubating the eggs. Some species guard their eggs until they hatch and allow the babies to stay with their mother for up to two years. Still other females hold their fertilized eggs inside their bodies until the eggs hatch, bearing live young.

CONTENT DEVELOPMENT

Remind students that the story about dragons in the text is just that—a story. There is a reptile on Earth today, how-

Although reptile sperm cells cannot be seen without a microscope, their egg cells are extremely large and visible to the unaided eye. The yellowish yolk in a reptile egg is actually the egg cell. Egg cells are ❶ immense because they contain huge amounts of stored food. After an egg is fertilized, the female's body builds a shell around it. In some reptiles, the female's body may add a layer of egg white (a thick protein-rich liquid) around the egg cell before the shell is formed. The egg white cushions the embryo, provides extra protein and water, and helps to prevent bacterial infections.

Most reptiles lay their eggs soon after the shells have formed. However, a few reptiles, such as certain snakes and lizards, protect their eggs by retaining them inside the body for part or all of the time it takes for the embryo to reach the stage at which it is ready to be hatched. In almost all reptiles in which the offspring are born alive, the developing young are nourished entirely by the yolk. In a small minority of live bearers, a special connection develops between the embryo's outermost membrane and the body of the female. Through this connection, food and oxygen are delivered from the female's body to her developing offspring. Reptile embryos with this special connection typically have much less yolk than embryos without this connection. Why do you think this is so? ❷

Whether they hatch from eggs or are born alive, young reptiles look like miniature adults. Unlike most amphibians, reptiles do not undergo metamorphosis. Because reptile eggs develop and hatch on land, there is no need for young reptiles to go through an immature water-dwelling stage. They can complete their development inside the safety of the eggshell. How does the more complete development of a reptile relate to the fact that reptile eggs contain a larger amount of yolk than typical amphibian eggs do? ❸

Lizards and Snakes

"Here be dragons" ancient maps sometimes declared about the faraway, poorly known lands at their edges. Of course, as travel to distant places increased, it soon became apparent that there were

ever, that closely resembles reptiles that lived during the dinosaur age. This reptile is a rare, nearly extinct lizard called a tuatara and it is of interest to paleontologists because it retains features of the ancient reptiles from which it evolved.

Have students observe the tuatara pictured in Figure 4–1. Point out that tuataras, which are found only on a few small islands off the coast of New Zealand, lead a leisurely life. Unlike many

no fire-breathing winged reptilian monsters anywhere in the world.

But as things turned out, there are dragons—of a sort. In the early part of this century—so the story goes—a pioneering pilot was trying to fly a plane from the islands of Indonesia to Australia. About 1500 kilometers east of Djakarta, the capital of Indonesia, the pilot developed engine trouble and was forced to land on a tiny volcanic island called Komodo. While the pilot tried to repair the engine, the dragons appeared and charged toward him.

The dragons were giant reptiles about 3 meters long and 160 kilograms in mass. They had scaly brownish hides, clawed feet, powerful tails, and short strong legs. Their long, forked tongues flickered in and out of their mouths like thin orange flames. The pilot probably did not wait to see the dragons' teeth before retreating to the safety of the cockpit.

Fortunately, the pilot was able to repair his engine. Although the dragons of Komodo usually hunt animals such as chickens, goats, small deer, and pigs, they are capable of killing large animals such as water buffaloes and humans.

The pilot's tale, along with similar stories from Indonesian pearl divers and fishers, prompted an expedition to obtain scientific specimens of the dragons. In 1912, the Komodo dragon was given its scientific name and correctly identified as a lizard—the largest (in terms of both length and mass) species of lizard in existence today.

Lizards are reptiles that typically have slender bodies, movable eyelids, long tails, four legs, and

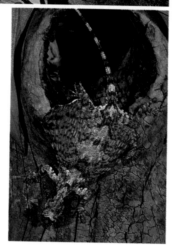

Figure 4–6 *Fire-breathing they are not, but all these lizards are known as dragons. The Komodo dragon (top left), water dragon (top right), and flying dragon (bottom) make their home in Southeast Asia. When fully spread, the reddish flaps of skin and ribs on the flying dragon's sides form "wings," which allow it to glide from tree to tree.*

reptiles, they are active at night, hunting the small animals they eat. In tuataras, the part of the brain called the pineal gland is located on the top of the skull in a place where several bones of the skull meet. This type of pineal gland is sometimes called the "third eye" because it contains cells that are sensitive to light. Tuataras, however, do not actually see with this pineal gland, or "third eye." Instead, they use it to detect changes in day length.

● ● ● ● **Integration** ● ● ● ●

Use the journey of the pioneering pilot to integrate social studies concepts into your science lesson.

ENRICHMENT

The reptilian muscle and skeletal systems exhibit many advances over those of amphibians. Compared to amphibians, reptiles with legs have larger, stronger limbs whose movements are well controlled. For example, the legs of many reptiles are well adapted to walking, burrowing, swimming, or climbing. Snakes, which lost their legs in the course of evolution, move by pressing large ventral scales against the ground. By expanding and contracting the muscles around their ribs in waves, snakes dig these ventral scales into the ground and push themselves forward. Because this kind of movement is slow and quiet, it makes snakes masters at stalking prey. Have interested students find out about the muscle and skeletal systems of other reptiles and report their findings to the class.

GECKOS

Geckos, like cats and dogs, are sometimes kept as house pets. At night, they catch and eat household pests such as ants, roaches, and spiders.

Figure 4–7 *A basilisk can run across the surface of small ponds and streams (bottom right). Geckos have suction-cuplike toes that enable them to walk up vertical panes of glass and run upside down across ceilings (top right). The horned lizard, which is often called a "horned toad," changes color. This helps it to blend in with its surroundings and to better absorb or reflect heat (left).*

clawed toes. They are placed in the same group of reptiles as snakes. Lizards range in size from tiny geckos 3 centimeters long to tree-dwelling monitor lizards of New Guinea that are more than 4.5 meters long. And as you can see in Figure 4–7, the shapes of lizards also vary.

For the most part, lizards are insect eaters that capture their prey by waiting for it to come nearby. When the prey is within range, the lizard lunges forward and grabs its meal in its jaws. Some lizards have evolved interesting variations on this maneuver. Slow-moving chameleons flick their long sticky tongue out of their mouth and then snap it back inside with a meal attached. The Gila (HEE-lah) monster of the American Southwest subdues its prey by poisoning it. After the Gila monster bites its prey, it hangs onto it tightly. A slow-acting poison made by glands in the lizard's lower jaw flows along grooves in the teeth and into the wounded prey. Contrary to popular stories, Gila monsters do not attack humans unless severely provoked. And although Gila monster bites are not deadly, they are extremely painful.

4–1 (continued)

CONTENT DEVELOPMENT

Remind students that the sizes and shapes of lizards vary. For the most part, lizards are insect eaters that capture their prey by waiting for it to come nearby. Some lizards lunge and use their jaws to catch prey; others use long, sticky tongues.

• **How does the poisonous Gila monster catch its prey?** (The Gila monster bites its victim and forces poison into the wound from a poison gland in the lower jaw. It then hangs tightly onto its prey, as the poison is slow-acting.)

Point out that Gila monsters do not generally attack humans unless severely provoked and that their bites are usually not fatal. Nevertheless, there are about eight known cases of humans being killed by the bite of a Gila monster. The only other venomous lizard is the beaded lizard of western Mexico.

CONTENT DEVELOPMENT

Ask questions similar to the following to help students understand the defense

mechanisms, or behaviors for survival, of lizards:

• **How do some types of lizards, such as the chameleon, protect themselves from attackers?** (The chameleon is one of several lizards that have the ability to change the color of their bodies to match their surroundings.)

• **How is the way a chameleon protects itself like that of a flounder?** (If students cannot recall their studies of flounder, lead them to suggest that both organisms

Figure 4–8 *Zap! A chameleon nabs its dinner with a flick of its long sticky tongue. How does the Gila monster catch and subdue its prey?* ●

Some lizards have special adaptations that help them to avoid becoming another animal's dinner. Chameleons are one of several kinds of lizards that can change color to match their surroundings. Other lizards have an even stranger way of protecting themselves. If threatened or captured by a predator, these lizards shed their tail. The castoff tail thrashes on the ground, confusing the predator and giving the lizard a chance to escape. Later the lizard grows back the missing tail.

Snakes are basically lizards that have lost their limbs, eyelids, and ears during the course of their ● evolution into burrowing forms more than 80 million years ago. Surprisingly, these losses did not restrict snakes to being burrowers forever. Snakes are an evolutionary success story—there are many species that live in many different kinds of habitats throughout the world.

A snake moves by wriggling its long, thin, muscular body. The scales on its belly help the animal to grip the surface on which it is moving and push itself forward. Special kinds of wriggling motions enable desert snakes to move across loose desert sand and allow tree-dwelling snakes to creep silently along branches. Many snakes are at home in the water and can swim at the surface or remain submerged. Sea snakes, which spend their entire lives in the ocean, have flattened, paddle-shaped tails that help them to swim swiftly.

Contrary to expressions such as "slimy as a snake," snakes are not at all slimy. They are actually rather pleasant to the touch—cool, dry, and smooth,

Figure 4–9 *In some lizards, the tail is structured in such a way that it can break off cleanly. Look closely and you can see the lizard's lost tail in the foreground. How do breakaway tails help lizards to survive?* ●

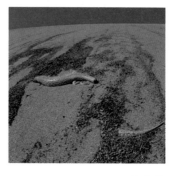

C ■ 97

● ● ● ● **Integration** ● ● ● ●

Use the discussion about the adaptation of burrowing to integrate concepts of evolution into your lesson.

GUIDED PRACTICE

▶ *Laboratory Manual*

Skills Development

Skill: Making observations

At this point you may want to have students complete the Chapter 4 Laboratory Investigation in the *Laboratory Manual* called Adaptations in Lizards. In this investigation students will observe the external structures of a lizard and determine the changes that occur in a lizard's coloration in different environments.

have the ability to change colors to better blend with their surroundings.)

• **Why is the ability to change color of benefit to an animal?** (The animal is less visible to attackers because its changed body color blends in with, or becomes a part of, the color of its surroundings.)

• **What is another defense mechanism possessed by some lizards?** (Some lizards possess the ability to discard their tail.)

• **What happens if predators attempt to catch these tail-shedding lizards by their tail?** (The tail becomes detached, possibly confusing the predator and allowing the lizard to escape.)

Remind students that several species of lizards (especially skinks) have this ability to shed the tail.

• **In what way is a lizard that sheds its tail somewhat like a starfish?** (Have students recall that a starfish can regenerate missing body parts. Therefore, both animals have the ability to regenerate missing body parts.)

Snake venoms can be deadly because they cause extensive damage to any living tissue they touch. Many years ago, people thought some snake venoms poisoned nerve tissue specifically, whereas other venoms caused the death of many different cell types.

New studies show that most venoms contain many toxic compounds that work in a variety of ways. All snake venoms contain powerful protein-digesting enzymes that cause massive tissue damage near the point of injection. Several venoms also block the transmission of nerve impulses in the prey. Many times the symptoms of snakebite depend on the location and the severity of the bite.

Figure 4–10 *Many snakes, such as the sea snake (top left) and the Siamese cobra (bottom left), have attractive colors and markings. The blue snake is a rare form of the green tree python (right).*

with little grooves around the edges of the scales. A word of warning, however: Snakes should be handled only under the supervision of an expert. Even tame nonpoisonous snakes can inflict painful bites.

Because snakes feed on small animals such as rats and mice, they can be quite helpful to people. But because some snakes are poisonous, people often try to get rid of all the snakes in an area. How would such an action affect the rat and mouse population in that area?

Snakes have a number of interesting adaptations for obtaining food. Although snakes are deaf and have poor eyesight, their other senses make up for these limitations. When a snake flicks its tongue in and out of its mouth, it is actually bringing particles in the air to a special sense organ on the roof of its mouth. This organ "analyzes" chemicals in the air, enabling the snake to find food. Many snakes are able to detect the body heat produced by their prey through special pits on the sides of their head.

Some snakes have glands in their upper jaw that produce a poison that immobilizes their prey. This poison is injected into the prey through special teeth called fangs. Four kinds of poisonous snakes make their home in the United States: rattlesnakes, copperheads, water moccasins, and coral snakes. Other poisonous snakes—such as the king cobra, which is the largest poisonous snake in the world—live on other continents.

4–1 (continued)

CONTENT DEVELOPMENT

Discuss the similarities and differences between lizards and snakes. Stress that although snakes have a bad reputation, only about 200 of the 2500 known species are poisonous. Snakes are actually more helpful than harmful because they kill a large number of rodents.

Explain that snakes have several adaptations that permit them to swallow prey larger in width than their own bodies. The lower jaw is not joined directly to the skull. During swallowing, the lower jaw can be drawn down and forward, allowing the snake to consume large prey. Therefore, the body can expand considerably during swallowing. All snakes

swallow their food whole, and all digestion takes place in the stomach.

Discuss some of the snake's adaptations for locating prey by asking questions similar to the following:
• **Look at the photograph of the snake sticking out its tongue in Figure 4–11. The holes are not the nostrils but sense organs called pits. What do the pits sense?** (Heat.)
• **Why do snakes flick their tongues in and out of their mouths?** (The tongue

picks up molecules from the air that are given off by nearby prey.)

Explain that the snake transfers the molecules from the air and its tongue to two cavities called Jacobson's organ, located in the roof of its mouth. The Jacobson's organ is extremely sensitive to odors. Also mention that snakes lack outer ears; consequently, they cannot hear in the same way that people and other animals hear. They do have inner ears, however, that can detect vibrations.

Turtles

Turtles are reptiles whose bodies are enclosed in a shell. The shell of a turtle consists of plates of bone covered by shields made of the same substance as scales. Some turtles have extremely strong shells that can support a weight 200 times greater than their own. This is roughly equivalent to your being able to hold two elephants on your back! Not all turtles have hard, bony shells. The leatherback sea turtle, the largest living turtle, has only a few small pieces of bone embedded in the skin of its back.

Turtles do not have teeth. Instead, they have beaks that are similar in structure to the beaks of birds. Many turtles eat plants as well as animals. The alligator snapping turtle has a particularly interesting adaptation for obtaining food. The turtle lies absolutely still on the bottom of a river or pond, looking like a rock or log. The only part of the turtle that moves is a small wormlike structure on the floor of its mouth. When a hungry fish swims into the turtle's mouth to eat the wriggling "worm," the turtle snaps its jaws closed and swallows the fish.

Figure 4–11 *Imagine trying to swallow something as big as your head! This action is impossible for humans, but snakes do it all the time. How do the tongue, teeth, and body muscles of snakes help them to obtain food?* ❷

ANNOTATION KEY

Answers

❶ The number of rats and mice would probably increase significantly. (Relating cause and effect)

❷ A snake's tongue helps it to sense food, its teeth inject poison, and its body muscles help to move swallowed food toward its stomach. (Relating concepts)

INTEGRATION

LITERATURE

For centuries, the image of the snake has been used in literature. Have interested students investigate some ancient folklore, myths, or legends to discover how the snake has been portrayed. Findings can be shared with the class.

ENRICHMENT

The world's largest snakes are the anaconda and the reticulated python. Both are reported to grow up to 10 meters in length, and both of these giant snakes kill their prey by constriction. There are occasional reports of these snakes attacking humans, but such incidents are rare. There is, however, a confirmed case of a reticulated python attacking and swallowing a 90-kilogram bear.

One of the world's smallest snakes is the Braminy blind snake. It is only about 15 centimeters in length.

REINFORCEMENT/RETEACHING

▶ *Activity Book*

Students who need practice on the concept of how reptiles move should be provided with the Chapter 4 activity called Reptile Locomotion. In this activity students will observe the structural adaptations that enable reptiles to move.

CONTENT DEVELOPMENT

Stress that turtles and tortoises are distinctly different from other reptiles in several ways. The most obvious difference is the bony shell that encloses the body. In most turtles, the shell is composed of a number of hard plates, but in soft-shelled species, it is composed of a leathery skin. The inner layer of the shell is usually fused with the vertebrae and ribs. Turtles and tortoises are also different in that they lack teeth. They catch and crush their food with horny beaks.

Figure 4–12 *Land-dwelling turtles with domed shells are often called tortoises (top left). The leatherback, the largest living species of turtle, leaves the ocean only to lay its eggs (top right). The desert tortoise uses its beak to nip off tasty bits of plants (bottom left). The matamata of South America spends most of its time hiding among the dead leaves and mud at the bottom of streams (bottom right). It feeds by sucking in water and unwary fishes like a vacuum cleaner.*

Sea turtles known as green turtles are among the most outstanding navigators in the animal kingdom. Soon after hatching, the young turtles head for the ocean. There they wander for many years over thousands of square kilometers. Eventually, the turtles mature and mate. Ready to lay their own eggs, these turtles do something quite amazing. They return to the same beach where they were born!

That beach may be hundreds of kilometers away, across an ocean surface that has no road signs or other markings. Yet the turtles find their way home. How? Recently, scientists have discovered that sea turtles are able to use wave motion and magnetic fields to maintain their direction.

Figure 4–13 *Sea turtles spend most of their lives at sea. But when they are ready to lay their eggs, the turtles return to the same beaches where they were hatched.*

100 ■ C

4–1 (continued)

CONTENT DEVELOPMENT

Write the terms *turtle, tortoise,* and *terrapin* on the chalkboard. Ask students to investigate the definition of each term.
• **What is the major difference between turtles and tortoises?** (Turtles spend most of their time in water, whereas tortoises spend most of their time on land.)
• **What is the meaning of the term terrapin?** (Strictly speaking, turtles live in the sea, and terrapins are freshwater turtles. Point out, however, that even biologists do not often make this distinction.)
Have students observe Figure 4–13. Ask:
• **In what way are the legs of turtles or terrapins adapted for living in water?** (They are shaped like paddles to facilitate swimming.)
Have students observe Figure 4–12. Ask:
• **In what way are the legs of the land-dwelling tortoises in these photo-**graphs different from the legs of turtles and terrapins? (Land-dwelling tortoises have stumpy legs for walking and claws on their feet for digging.)

ENRICHMENT

Students might be interested to learn that turtles are the oldest group of living reptiles. The fossil record shows that their ancestors were around even before the dinosaurs, and they have changed little throughout time. Turtles are, how-ever, a small group; there are only a few more than 200 species, several of which are threatened with extinction.

Turtles and tortoises have a life expectancy greater than that of most other vertebrates. Most species of turtles live 50 years or more, with the record age of any turtle being about 150 years.

Alligators and Crocodiles

Alligators and crocodiles are large meat-eating lizardlike reptiles that spend much of their time in water. They have long snouts, powerful tails, and thick, armored skin. Although alligators and crocodiles are similar, it is not difficult to tell them apart. Alligators have broad, rounded snouts, whereas crocodiles have narrow, pointed snouts. When an alligator's mouth is closed, only a few of the teeth on its lower jaw are visible. When a crocodile's mouth is closed, most of its teeth are visible. But don't be taken in by the crocodile's welcoming grin. Crocodiles are far more aggressive than alligators, and some species are known to eat humans!

When they are not lying on riverbanks basking in the sun or resting in the shade, alligators and crocodiles spend their time submerged in water. Although these reptiles look lazy and slow, they are capable of moving rapidly, both in water and on land.

Alligators and crocodiles do most of their hunting at night. They eat everything from insects, fishes, and amphibians to birds and large hoofed mammals. (Larger alligators and crocodiles typically hunt larger prey.)

Alligators and crocodiles build nests of mud or plants in which they lay their hard-shelled eggs. In some species, the eggs are abandoned after they are laid. But in other species, the female takes good care of her eggs and offspring.

Figure 4–14 *After her eggs hatch, a female alligator or crocodile will carry her babies in her jaws. The female will continue to care for her young—often with the help of the male.*

Figure 4–15 *Alligators have broad, rounded snouts (bottom left). Crocodiles have narrow, pointed snouts and a distinctive toothy "grin" (bottom right). The gharial of India belongs to the same group of reptiles as alligators and crocodiles (top).* ①

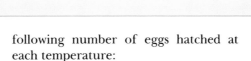

C ■ 101

HISTORICAL NOTE
GALAPAGOS TORTOISES

The great tortoises of the Galapagos Islands have been so reduced in number that they are now in danger of extinction. When the islands were first discovered in the seventeenth century, the giant tortoises were reported to be so numerous that a person could walk long distances atop their shells without touching the ground. In the 1800s, sailors started catching them for their meat and oil. Later, the dogs and cats of settlers further reduced the population when they started eating the eggs and babies of the tortoises.

clude that temperature is a factor in determining the sex of green turtles.)

Point out that actual research has confirmed that when green turtle eggs are incubated at low temperatures, mostly males hatch. Higher temperatures produce mostly females. Similar studies have shown that temperature variations affect sex determination in other reptiles, too.

GUIDED PRACTICE

Skills Development

Skill: Drawing conclusions

After discussing the remarkable navigational abilities of the green turtle, present this hypothetical experiment to the class: A biologist collected a large quantity of green turtle eggs from a beach and placed them in groups of 25 each in incubators of various temperatures. After approximately 60 days, the following number of eggs hatched at each temperature:

26°C: 21 males, 2 females
27°C: 18 males, 0 females
28°C: 13 males, 11 females
29°C: 12 males, 12 females
30°C: 1 male, 19 females
31°C: 0 males, 21 females
32°C: 1 male, 20 females

• **What might be concluded from the biologist's data?** (Students should con-

REINFORCEMENT/RETEACHING

Students sometimes have difficulty determining common traits of reptiles. To better understand the characteristics of reptiles, have students consider the following true or false statements:

• **All reptiles hatch from eggs.** (False.)
• **Reptiles spend part of their life on land and part of their life in water.** (False.)
• **Reptiles reproduce by external fertilization.** (False.)

PROBLEM SOLVING

ALLIGATOR ANXIETIES

This feature enables students to reinforce the idea that temperature variations affect sex determination in reptiles.
1. The hypothesis that individual students develop to explain the results of the project will vary, depending on the student. Most students, however, will probably suggest that higher incubation temperatures tend to develop males and lower incubation temperatures tend to develop females.
2. In their experiments, students should expect to prove their hypothesis that temperatures are related to sex determination in alligators.
3. If the various hypotheses prove correct, conservation officials could use the results to influence male and female birth rates in captive breeding programs.

Alligator Anxieties

The purely imaginary Gatorville Amateur Conservationist Society (GACS) is faced with a puzzling situation. Three months ago, the Gatorville swamp was drained to make way for a new shopping mall. The adult alligators in the swamp were moved to a wildlife refuge elsewhere in the state. GACS volunteers rescued the newly laid eggs from the alligator nests and placed them in specially designed incubators.

The temperature of all the incubators was set at 30°C. One incubator, however, had a faulty thermostat. The actual temperature in this incubator was 4°C higher than what was indicated on the dials. This problem was not discovered until the eggs had been in the incubators for three weeks.

To everyone's delight, most of the rescued alligator eggs did hatch. But an examination of the baby alligators revealed something strange: All of the

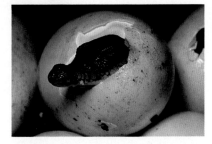

babies from the normal incubators were females, and all the babies from the faulty incubator were males. Why?

Designing an Experiment
1. Develop a hypothesis to explain the results of the alligator-hatching project.
2. Design an experiment to test your hypothesis. What do you expect the results of your experiment to be?
3. If your hypothesis proves to be correct, how might this information affect future conservation efforts?

4–1 Section Review

1. Describe three ways in which reptiles are adapted for life on land.
2. How does the structure of the egg help it to perform its function?
3. Compare the three major groups of reptiles.

Connection—*Wildlife Conservation*
4. Conservationists are concerned, that once sea turtles that nest on a particular beach are killed off, there will never be sea turtles on that beach again. Explain why.

102 ■ C

4–1 (continued)

ENRICHMENT

▶ *Activity Book*

Students will be challenged by the Chapter 4 activity in the *Activity Book* called Investigating How Reptiles Regulate the Temperatures of Their Bodies. In this activity students will determine how a reptile's behavior helps to maintain its body temperature.

INDEPENDENT PRACTICE

Section Review 4–1

1. Land adaptations for reptiles include lungs for breathing, scaly, waterproof skin to prevent water loss, and eggs with protective coverings to prevent embryos from drying out.
2. The structure of an egg contains all the nutrients that a growing embryo will need and a covering or shell that protects the embryo and prevents the contents of the egg from drying out.

3. All the reptile groups are similar in that they are coldblooded vertebrates that have lungs, scaly skins, and special types of eggs that are laid in a dry environment.
4. Accept logical answers.

REINFORCEMENT/RETEACHING

Monitor students' responses to the Section Review questions. If students appear to have difficulty with any of the questions, review the appropriate material in the section.

4-2 Birds

People's admiration and affection for birds are reflected in the frequent use of birds as symbols. The eagle shows up on the back of quarters and on postage stamps as the national emblem of the United States. Certain airlines and other businesses feature birds in their emblems. Even sports teams are named after birds. The Toronto Blue Jays, Seattle Seahawks, and Phoenix Cardinals are just three such teams. Can you name others?

Many people think that birds are the most fascinating and colorful animals on Earth. One reason for this is that birds can fly. Along with bats and insects, birds are the only animals with the power of flight, although not all species of birds fly.

Birds are relatively recent additions to the parade of life. Because the skeletons of many small dinosaurs are almost identical to the skeletons of the earliest birds, there is much controversy over which fossils are those of birds and when birds first appeared on Earth. The oldest fossil that is definitely that of a bird is of *Archaeopteryx* (ahr-kee-AHP-ter-ihks). The root word *archaeo-* means ancient, and the root word *-pteryx* means wing. Why is *Archaeopteryx* an appropriate name? ❶

Archaeopteryx lived about 140 million years ago, during a time when dinosaurs and other reptiles ruled the Earth. As you can see in Figure 4–16, this bird did not look much like modern birds. It had a long bony tail and sharp teeth, neither of which is found in modern birds. It had clawed fingers and many other odd features not typical of birds. Despite all the ways in which it was different from modern birds, *Archaeopteryx* was definitely a bird. For around its fossilized bones are the unmistakable impressions of feathers.

Birds are warmblooded egg-laying vertebrates that have feathers. All modern birds—including ostriches, penguins, and other flightless birds—evolved from ancestors that could fly. As you read about birds, focus on the ways in which their characteristics reflect their heritage of flight.

The single most important characteristic of birds is **feathers**. Feathers, like the scales of reptiles, are

Guide for Reading

Focus on these questions as you read.

▶ What are the major characteristics of birds?

▶ How are birds adapted for flying?

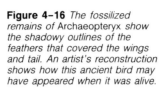

Figure 4–16 *The fossilized remains of* Archaeopteryx *show the shadowy outlines of the feathers that covered the wings and tail. An artist's reconstruction shows how this ancient bird may have appeared when it was alive.*

C ■ 103

MULTICULTURAL OPPORTUNITY 4-2

Birds are often featured on Chinese decorative objects such as vases, boxes, trays, screeens, and fans. Sometimes they are real; other times they are imaginary. Chinese tapestries, woven in silk and gold threads, frequently include the mythical bird *feng huang*, which is considered to be the king of birds and the symbol of happiness. Real birds often depicted by Chinese artists include cranes, pheasants, peacocks, ducks, and egrets. Interested students may want to look for examples of Chinese art in magazines and on postcards. They also may want to research Chinese proverbs and sayings that are associated with birds.

ESL STRATEGY 4-2

Point out that *communicate, migrate,* and *navigate* are terms describing special abilities that birds possess. After making sure students understand the words, have students write three short paragraphs explaining how birds are able to perform these complex behaviors. Volunteers can read their paragraphs to the class.

Give students labeled pictures of 20 different birds that are to be classified as birds of prey, flightless birds, songbirds, or waterfowl. Have students use these categories as headings for a chart, listing each bird in alphabetical order under its proper group, or heading.

tebrates that have wings and a body covered with feathers.

● ● ● ● **Integration** ● ● ● ●

Use the names of professional athletic teams to integrate physical education concepts into your science lesson.

Use the derivation of the term *Archaeopteryx* to integrate language arts concepts into your science lesson.

Use the description of birds of the early Earth to integrate concepts of evolution into your lesson.

CLOSURE

▶ *Review and Reinforcement Guide*

At this point have students complete Section 4–1 in the *Review and Reinforcement Guide.*

TEACHING STRATEGY 4-2

FOCUS/MOTIVATION

📽 **Media and Technology**

Have students examine the relationship between wetland ducks and their environment through the use of the Interactive Videodisc/CD ROM entitled Paul ParkRanger and the Disappearing Ducks. Students will begin their exploration in Paul ParkRanger's cabin where they will find helpful clues and interesting surprises.

CONTENT DEVELOPMENT

Point out that birds are members of the phylum Chordata. Explain that birds make up the class of warmblooded ver-

Discovery Learning

Skills: Making observations, making comparisons

Materials: contour feather, down feather

Students will discover that most feathers have a stiff central shaft, on each side of which is a flat vane. The vane is made up of hundreds of thin, parallel branches from the shaft. These branches are called barbs. The largest feathers are the long contour feathers. These feathers give birds their streamlined shape, which helps them when flying. In addition to feathers with vanes, some birds have down feathers. Down feathers have a short shaft and soft, fuzzy barbs that are not connected into vanes. The down feathers provide insulation.

Figure 4–17 *Body feathers help to insulate the mourning dove, and feathers on its wings and tail help it to fly (left). The brightly colored feathers of the rainbow lorikeets help them to communicate (right). Feathers also hide birds from predators. Can you locate the four ptarmigans (center)?*

Activity Bank

Do Oil and Water Mix?, p.163

ACTIVITY
DISCOVERING

Comparing Feathers

Obtain a few samples of feathers. Try to obtain both down and contour feathers. How are the feathers similar? How are they different?

Gently run your fingers along a contour feather from its tip to its base. What happens? What happens if you rub the feather in the opposite direction?

■ How are feathers put together?

■ How does the structure of a feather help it to perform its functions?

made of dead cells that contain the same material found in your fingernails. Feathers come in many colors, shapes, and sizes. Body feathers help to insulate the body. Feathers on the wings and tail are used in flying. Dull-colored, speckled feathers may help a bird to blend in with its background, hiding it from its natural enemies. Brightly colored feathers help a bird to communicate with other members of its species. For example, the brilliant feathers of male birds such as peacocks advertise their presence to potential mates.

The feathers on the wings and on most of a bird's body are called **contour feathers.** Contour feathers are the largest and most familiar feathers. They give birds their streamlined shape. Other feathers—called **down**—are short, fluffy feathers that act as insulation. Most birds have down feathers on their breasts. As you can see in Figure 4–18, baby birds are often covered with down. As the baby birds grow up, contour feathers grow in and most of the down falls out. Why do you think down from birds such as geese is often used in coats and quilts?❶

Have you ever heard the expression "eats like a bird"? This phrase, which is used to describe someone who eats very little, was certainly not invented by someone familiar with birds. For birds "eat like pigs." In fact, birds are even bigger eaters than pigs. Because they are warmblooded, birds must expend energy in order to maintain their body temperature. Flying also demands great amounts of energy. In order to meet these energy demands, birds must acquire a lot of energy in the form of food. A

❶

4–2 (continued)

CONTENT DEVELOPMENT

Have students recall that warmblooded organisms can maintain a constant body temperature despite the temperature of the environment. Point out that many birds maintain a higher body temperature than that of most other vertebrates.

CONTENT DEVELOPMENT

Explain that feathers grow from little pits in a bird's skin. Point out that there are four kinds of feathers that grow on a bird. The soft down feather is the first type of feather a bird grows. It is easily seen on baby birds (refer students to Figure 4–19), growing close to the skin and acting as insulation.

• **What are some reasons people own or purchase a down jacket?** (Accept all answers. Students might suggest that down jackets are warm and lightweight.)

• **Why is down a good insulator?** (Lead students to suggest that down "puffs" or "lofts," allowing air to be trapped inside the tiny fibers.)

Explain that next to the down feathers are filoplumes. Filoplumes are hairlike feathers with a small tuft on the end.

Point out that the contour feather is the most visible feather on a bird. Contour feathers streamline a bird's body, helping the flight of the bird and also providing color.

Tell students that large, strong quill feathers are in the wings and tail of a bird. They aid the flight of a bird by giving the bird the ability to maneuver. (These feathers are analogous to the adjustable parts of an airplane tail and airplane sidewings—maneuverability is provided.) These feathers of flying birds form an airfoil that allows the bird to lift from a surface. When a bird flaps its wings, pressure is increased on the underside of the wing and pressure is re-

pigeon eats about 6.5 percent of its body weight in seeds, crumbs, and other foods every day. A hummingbird eats about twice its weight in nectar (a sugary liquid produced by flowers) daily. How much food would you have to eat every day if you ate like a pigeon? If you ate like a hummingbird? ②

Birds eat many different kinds of foods, including microscopic blue-green bacteria, fruits and seeds, insects, other birds, mice, monkeys, and the remains of dead animals. The beak of a bird is often remarkably adapted for the type of food it eats. Hawks and owls have sharp, curved beaks used for tearing their prey into pieces small enough to be swallowed. Kiwis and woodcocks have long, thin beaks that are used to probe into the soil for earthworms. Cardinals and sparrows have thick blunt beaks with which they crush the hard shells of seeds.

Bird adaptations for flight are not limited to the outside of the body. Bird bones are hollow and

Figure 4–18 Birds have fluffy down feathers and leaf-shaped contour feathers. As this young frigate bird grows older, most of its whitish baby down will be replaced by glossy black adult feathers.

Figure 4–19 You might think that a toucan's enormous beak would cause the bird to tip over. But the beak is hollow and light for its size (top left). The puffin's tall, flat beak serves as a shovel for digging nesting holes. It is also a handy fish holder (top right). The pelican's pouched beak scoops fishes from the water like a net (bottom).

BACKGROUND INFORMATION

FILOPLUMES

Another type of feather, called the filoplume, has a long, slender quill with only a few small barbules at the tip. For a long time, biologists thought filoplumes had no function. It is known now that these feathers carry many sensory cells that detect vibrations in nearby feathers. It is believed that filoplumes help birds keep their feathers properly adjusted during flight and other activities.

duced on the upperside of the wing. As a result, air spills around the tip of the wing in an upward direction to reduce this difference. The result is upward lift.

ENRICHMENT

Ask each student to compile a list of five bird traits. For example, students might suggest that birds (1) are usually capable of flight, (2) have feathers, (3) are warmblooded, (4) lack teeth, and (5) have scales on their legs.

After individual lists have been compiled, have a volunteer compile a master list on the chalkboard. Discuss each trait as it is listed.

CONTENT DEVELOPMENT

Ask students to comment on the often-used phrase "He or she eats like a bird." Point out that many people erroneously believe that birds do not eat very much. Birds eat more for their size, however, than any other vertebrates, do.

• **Why do you think birds require so much food?** (Answers will vary, but develop the idea with students that birds must consume great amounts of food to maintain the high metabolic rate that is required for flight.)

● ● ● ●   Integration   ● ● ● ●

Use the expression "eats like a bird" to integrate language arts concepts into your science lesson.

Several birds can see in what is the ultraviolet part of the spectrum, and many can see well into the infrared region. In addition, the complex construction of their retinas undoubtedly enables many birds to see more subtle shades of colors than we can distinguish. Bird's retinas also enable them to see clearly in three places at once. Human retinas, in contrast, enable us to look closely in only one place at a time.

Figure 4–20 *Some birds store food. A shrike kills prey and then hangs them on thorns and twigs. This shrike looks quite pleased about the striped caterpillar it has caught. How does an acorn woodpecker store food?* ❷

Activity Bank

An Eggs-aggeration, p.164

therefore quite lightweight. Like the bones, the internal organs of birds have evolved in ways that enable birds to fly efficiently. The respiratory system is more advanced in birds than in any other class of vertebrates. How does this relate to the energy needs of birds? (*Hint:* Oxygen is needed to break down food to obtain energy.) ❶

Birds have special structures called air sacs attached to their lungs. The air sacs inflate and deflate in a complex way to ensure that a supply of fresh air is constantly moving in one direction through the lungs. The great efficiency with which birds perform the function of gas exchange is particularly apparent at high altitudes, where the air is thin. Mountain climbers on Mount Everest must carry tanks of oxygen and stop often to rest. Their lungs cannot obtain oxygen efficiently enough to permit long periods of strenuous activity. As the mountain climbers rest, they may observe Himalayan geese flying overhead. The lungs of geese can obtain oxygen efficiently enough to permit the extremely strenuous task of flying. The geese even have spare air—which they use to honk a greeting as they soar past the out-of-breath mountain climbers!

Birds have a double-loop circulatory system in which the two loops are completely separated. This helps to ensure that the oxygen provided by the lungs is delivered effectively to the cells of the body.

Like reptiles and amphibians, birds have two long, oval kidneys that filter nitrogen-containing wastes from the blood. Birds produce the same kind of concentrated nitrogen-containing waste product as land-dwelling reptiles.

Although the term "bird-brained" means stupid, birds are actually quite intelligent. Like the rest of their nervous system, their brain is well developed. The eyesight of many birds—hawks, vultures, and eagles, to name a few—is far keener than that of humans. Some birds have an extraordinarily sharp sense of hearing. The faintest rustle of a mouse as it creeps across the forest floor is all that a hunting owl needs to pinpoint the location of its next meal.

The reproductive organs in birds are often tiny and compact. Only during the breeding season do these organs enlarge to a functional size. All female birds lay eggs. No birds, past or present, bear their

---

4–2 (continued)

CONTENT DEVELOPMENT

Point out that the respiratory system of birds is extremely efficient at taking in oxygen and eliminating carbon dioxide. Remind students that this should not be surprising because the high metabolic rate of birds demands an efficient gas-exchange system.

Explain that the reason for this efficiency is that bird lungs are connected at both the anterior and posterior to large air sacs in the body cavity and bones. When a bird inhales, air travels through passageways that lead into the lungs. Some of this air remains in the lungs, where gas exchange occurs. Most of the air, however, goes through the lungs into posterior air sacs. When a bird exhales, air from the posterior air sacs passes into the lungs for gas exchange. Stress that for this reason, birds are able to remove oxygen from the air when they inhale as well as when they exhale. Use students'

knowledge of their own respiratory system to confirm that gas exchange is more efficient than that in other animals. The air sacs of a bird serve an additional function: They make a bird's body more buoyant, allowing it to fly more easily.

● ● ● ● **Integration** ● ● ● ●

Use the discussion concerning the changes in the internal organ structure of a bird over time to integrate concepts of evolution into your lesson.

ENRICHMENT

You can make an interesting point to your students about the efficiency of birds' lungs by pointing out birds' responses to situations in which the availability of oxygen is limited. You can point out, for example, that many people who live at sea level get headaches and become weak when they visit cities such as Denver or other places located high up in the mountains. They have this reaction because as atmospheric pressure drops

young alive. How do these two characteristics of bird reproduction affect flying ability? ❸

Bird eggs, which have the same basic structure as reptile eggs, contain a generous supply of egg white and are covered by a hard shell. Because bird eggs will develop only if they are kept at the proper temperature, almost all are cared for by the parent birds. In some species, only one parent cares for the eggs. In other species, both parents take turns keeping the eggs warm.

In almost all birds, the parents' duties are not finished when the eggs hatch. In some birds, such as chickens and ducks, the young have feathers when they hatch, and they are soon able to run about and feed themselves. However, they still depend on their parents for protection. In other birds, the young are featherless, blind, and helpless when they hatch. Their parents must feed and care for them until they are old enough to fly and take care of themselves.

Bird Behavior

Have you ever heard a bird singing cheerfully on a fine spring day? If you have, you might have wondered what all the excitement was about. Perhaps you might have thought that the bird was happy.

Figure 4–21 *Most birds, such as the skimmer, crouch over their eggs to keep them warm (bottom). An Adelie penguin keeps its egg warm by balancing it on its feet (top).*

Figure 4–22 *Some birds are well-developed when they hatch. The newly hatched Canada geese will soon be waddling through the grass and swimming in the water—with a little encouragement from their parent! Others are featherless, blind, and helpless when they hatch. These newly hatched pelicans will be completely dependent on their parents for quite some time.*

Activity Bank

Strictly for the Birds, p.165

C ■ 107

at high altitudes, so does the partial pressure of oxygen. Low oxygen pressure, in turn, causes oxygen to diffuse into the air sacs of the lungs more slowly than it does at sea level. Have students discuss any altitude experiences they or their family members may have had.

Students may be interested to learn that to test the relative efficiency of bird and mammal lungs in simulated high-altitude conditions, a researcher experimentally exposed both sparrows and mice to the lowered atmospheric pressure found at 2000 meters above sea level. The sparrows felt fine and could even fly normally. The mice, on the other hand, became so weak they could barely move.

CONTENT DEVELOPMENT

Remind students again that despite the derogatory term "birdbrain," birds are rather intelligent animals. The brain of a bird is actually quite large, controlling behaviors such as flying, nest building, care of young, courtship, and mating.

An intelligent brain is also related to the development of senses. Birds also have extraordinarily well-developed eyes. Birds can easily see small objects and, in many cases, can see color better than humans can. Most species of birds can hear well, although their senses of taste and smell are not well developed.

Some species of birds migrate tremendous distances in search of warm winter homes. Challenge interested students to discover the distances that different species of birds migrate annually. Also challenge them to find the average weight of each of the bird species. Then have them compute to determine which species of bird flies farthest per gram of body weight.

Figure 4–23 *Male birds have many ways of attracting a mate. A peacock spreads the long, colorful plumes of his tail and struts about. The male frigate bird puffs up his bright red throat sac. Male weaverbirds demonstrate their skills at nest-building.*

Figure 4–24 *The meadowlark has one of the loveliest songs in nature. Why do birds sing?* ❶

Birds actually sing for rather serious reasons. Early in the breeding season, birds sing to attract a mate and to warn other birds of the same sex to stay away. The song establishes or maintains a **territory.** A territory is an area where an individual bird (or any other animal) lives. Establishing a territory is important because it ensures that fewer birds will compete for food and living space in the same area. Birds may also sing to warn of danger, to threaten an enemy, or to communicate other sorts of information.

Birds communicate with signals that are seen as well as with signals that are heard. The bright feathers of some male birds are used to attract females and to scare off rivals. However, brightly colored feathers also make the bird more noticeable to predators. Can you explain why many birds sport bright colors only during the breeding season? ❷

Some birds attract a mate by doing something unusual. Male bowerbirds build large and colorful constructions of twigs to make females notice them. They may even paint their bowers with berry juice or

---

4–2 (continued)

GUIDED PRACTICE

Skills Development

Skills: Making observations, applying concepts

At this point have students complete the in-text Chapter 4 Laboratory Investigation: Owl Pellets. In this investigation students will identify the different organisms that owls eat.

CONTENT DEVELOPMENT

Point out that birds, like all living things, must fight for survival. Birds establish a territory so that there will not be too many birds competing for food or living space in the same area. One way this is accomplished is by singing songs. The song a bird sings is sometimes sung to establish or maintain a territory.

Birds also sing to warn of danger, threaten an enemy, communicate other kinds of information, or court. Empha-

size the varied courtship behaviors of birds. Point out that in most cases these behaviors are instinctive.

• **Why are the feathers of many male birds more attractive and colorful than the feathers of a female?** (The bright plumage serves to attract females.)

Explain that the courtship displays of many male birds are quite elaborate. Most students are probably familiar with the male peacock's brilliant plumage and its strutting behavior. Birds of paradise

decorate them with feathers, shells, butterfly wings, and flowers. Male weaverbirds construct a nest that is examined by a prospective mate for soundness and craftsmanship. A male penguin does not construct anything. Instead, he presents his intended mate with a pebble. The pebble indicates that he is ready to breed and to care for a youngster.

Most birds build nests, which are designed to protect the eggs and the young birds as they develop. These nests can be little more than a shallow trench hollowed out in the ground or they can be quite elaborate. Hummingbird nests are tiny cups woven out of spider silk, decorated with bits of plants, and lined with feathers.

Many birds **migrate,** or move to a new environment during the course of a year. Some birds migrate over tremendous distances. For example, the American golden plover flies more than 25,000 kilometers when it migrates. Birds migrate for many reasons, but probably the most important reason is to follow seasonal food supplies. Birds have developed extremely accurate mechanisms for migrating. Scientists have learned that some birds navigate by observing the sun and other stars. Other birds follow coastlines or mountain ranges. Still other birds are believed to have magnetic centers in their brains. These centers act as a compass does to help the bird find its way.

Rara Aves

This Latin phrase, which literally means rare birds, is used to refer to any type of extraordinary thing. There are a number of works of literature that feature birds in their titles.

A few of these *rara aves* are *The Trumpet of the Swan* by E. B. White; *Jonathan Livingston Seagull* by Richard Bach; *To Kill a Mockingbird* by Harper Lee, and *I Know Why the Caged Bird Sings* by Maya Angelou.

Figure 4–25 *The nests of barn swallows are round clay pots built one beakful of mud at a time. The nests of certain African weaverbirds have several "rooms." The nest mounds built by mallee fowl are up to 4.5 meters high and 10 meters across.*

C ■ 109

also have brilliant plumage, and the males of some species carry out their courtship rituals while hanging upside down from trees. Have interested students research courtship behaviors of different birds and report their findings to the class.

● ● ● ● **Integration** ● ● ● ●

Use the description of a bird's competition for food and living space to integrate concepts of ecology into your lesson.

BIRD MUSCLES

To power the downward stroke necessary for flight, birds have large chest muscles. These muscles attach to a long keel that runs down the front of an enlarged breastbone, or sternum. The sternum, in turn, is firmly attached to the rib cage. In strong flying birds, such as pigeons, the chest muscles may account for as much as 30 percent of the animal's mass.

ANNOTATION KEY

Integration

1 Life Science: Classification. See *Parade of Life: Monerans, Protists, Fungi, and Plants*, Chapter 1.

2 Life Science: Evolution. See *Evolution: Change Over Time*, Chapter 2.

4–2 (continued)

REINFORCEMENT/RETEACHING

Remind students that nests are designed to protect the eggs and young birds as they develop. Help students recall that the female egg is fertilized by the male sperm "inside" the body and that the name for this process is internal fertilization. Point out that after the eggs are laid, the female bird usually incubates, or warms the eggs, by sitting on them. During this time, the male of the species often feeds the female.

Also remind students that after hatching, some birds are too small to hunt for food and water. Their parents care for them until they are old enough to take care of themselves.

CONTENT DEVELOPMENT

Point out that there are approximately 8700 living species of birds belonging to more than 160 families. Because of the difficulty in grouping this large, diverse

Types of Birds

There are about 8700 living species of birds. These species are divided among roughly 30 scientific classification groups. As you might imagine, such diversity among birds makes it impossible to discuss all the scientific groups in detail. To simplify matters, birds are often divided into a few broad categories according to one or two significant characteristics. Although these nonscientific categories give no information about evolutionary relationships and also exclude a number of birds, they do provide a glimpse of the enormous diversity among birds.

Most familiar birds are commonly known as songbirds. Cardinals, sparrows, and robins are examples of songbirds. Songbirds range in size from tiny flycatchers 8 centimeters long to birds of paradise whose elegant tail feathers make them more than a meter long. Songbirds have feet that are well adapted for perching on branches, electric wires, and other narrow horizontal structures.

As their name indicates, many songbirds sing beautifully. Nightingales, mockingbirds, warblers, and canaries make some of the loveliest sounds in the natural world. However, some birds with ugly voices—crows and ravens, for example—are also

Figure 4–26 *Songbirds, more correctly known as perching birds, include the scarlet-chested sunbird of Africa (top), the robin of Europe (bottom left), and the Gouldian finch of Australia (bottom right).*

110 ■ C

group of organisms, birds are often divided into broad, nonscientific categories. Write these four categories of birds—perching birds, birds of prey, waterfowl, and flightless birds—on the chalkboard.

Students might be interested to learn that there were even more kinds of birds in the past. Paleontologists estimate that more than 100,000 species of birds have become extinct since the Jurassic Period of our history.

● ● ● ● **Integration** ● ● ● ●

Use the different characteristics and categories of birds to integrate classification concepts into your lesson.

GUIDED PRACTICE

▶ *Laboratory Manual*

Skills Development

Skill: Making observations

At this point you may want to have students complete the Chapter 4 Labora-

placed in this group. Thus songbirds are more appropriately known as perching birds.

Hunting birds such as hawks, eagles, and owls are known as birds of prey. Birds of prey are superb fliers with keen eyesight. Soaring high in the air, they can spot prey on the ground or in the water far below them. Birds of prey eat fishes, reptiles, mammals, and other birds. Some even eat small monkeys.

Birds of prey are able to fly very fast. The peregrine falcon has been clocked at more than 125 kilometers per hour while diving at its prey. Birds of prey have sharp claws, called talons, on their toes. Talons enable the bird to grasp its prey. Some eagles have talons that are longer than the fangs of a lion. Birds of prey also have strong, curved beaks that are used to tear their prey into pieces small enough to be swallowed.

The birds that swim and dive in lakes and ponds are known as waterfowl. Swans and ducks are typical waterfowl. These birds glide across the surface of the water, propelled by webbed feet that resemble paddles. Occasionally, they duck their head and neck into the water to nibble at water plants with their broad, flat beak.

During the course of evolution, some birds have lost their ability to fly. The wings of these birds are

Figure 4-27 *The king vulture of South America (bottom left), the monkey-eating eagle of the Philippines (bottom right), and the bald eagle of North America (top) are examples of birds of prey.*

C ■ 111

MIGRATION NAVIGATION

Some migratory birds use a magnetic sense to navigate. This magnetic sense, located somewhere in the head, operates like a built-in compass, responding to the Earth's magnetic field. Because many birds can hear well, some migratory birds can navigate by sound, hearing for example the pounding waves on a shoreline when they are many kilometers away. Many other migratory birds use a combination of keen eyesight, instinct, and a built-in clock to navigate by the sun and stars.

and cassowary in Figure 4-29 differ from those of other birds on these pages? (They are adapted for rapid running.)

• The birds in Figure 4-28 have webbed feet and swim and dive in water. To which group do these birds belong? (Waterfowl.)

Remind students that though we usually think of birds as being able to fly, some species of birds cannot fly and spend their entire lives on the ground.

• • • • Integration • • • •

Use the example of birds losing their ability to fly to integrate concepts of evolution into your lesson.

INDEPENDENT PRACTICE

▶ *Activity Book*

Students who need practice on the concept of coldblooded vertebrate characteristics should complete the chapter activity Fact or Fiction? In this activity students will identify various statements about coldblooded vertebrates as fact or fiction. Have students relate these statements to birds as well.

tory Investigation in the *Laboratory Manual* called Examining Bird Adaptations. In this investigation students will examine some general characteristics of birds.

CONTENT DEVELOPMENT

Ask the following questions to help students identify the traits of the four bird groups:

• **What are four nonscientific groups of birds?** (Perching birds, birds of prey, waterfowl, and flightless birds.)

• **What characteristic identifies the songbirds in Figure 4-26 as perching birds?** (Students should notice that the feet are adapted for gripping branches.)

• **What are some other types of perching birds?** (Some familiar examples are cardinals and sparrows.)

• **What characteristic identifies the birds in Figure 4-27 as birds of prey?** (Strong, curved beaks and sharp claws called talons.)

• **How do the legs and feet of the ostrich**

4-2 (continued)

CONTENT DEVELOPMENT

Explain that a number of birds have lost the ability to fly. Some species, such as ostriches, spend their time walking or running on a powerful pair of hind legs. Their feet usually have three strong toes that make contact with the ground. Their wings are usually quite reduced in size and are incapable of lifting them off the ground. These birds can get quite large, as they have no need to minimize their mass. Still other birds have given up flying in favor of swimming. Their wings, legs, and feet are so reduced in size that they look quite comical on land. In water, however, their feet and wings are powerful flippers that enable them to "fly" through the water.

INDEPENDENT PRACTICE

▶ *Activity Book*

Students who need practice on the concept of characteristics of cold-blooded vertebrates should complete the chapter activity Unusual Coldblooded Vertebrates. In this activity students will identify characteristics of specific cold-blooded animals. Have students com-

pare these characteristics to the characteristics of birds.

ENRICHMENT

If possible, take your class on an early morning bird-watching tour of a wooded area. Students may use binoculars and field guides to help them identify birds. Artistic students may make and display colored sketches of the birds they observe. Students interested in photography may take and exhibit snapshots for the class.

Figure 4-28 *Swans (left) and geese (right) are types of waterfowl. What are the main characteristics of waterfowl?* ❶

small compared to the size of their body. Some flightless birds are specialized as runners. Such birds include the ostrich of Africa (the largest bird alive today), the rhea of South America, and the emu and cassowary of Australia. These birds have strong leg muscles that enable them to run quickly and to defend themselves against any enemy foolish enough to challenge them. Penguins are flightless birds that are specialized as swimmers. On land, penguins waddle and hop awkwardly. But in water, they are swift and graceful.

Figure 4-29 *The cassowary of New Guinea (top) and the ostrich of Africa (bottom right) have long, powerful legs that enable them to run swiftly. The comical parade of king penguins shows that penguins cannot move quickly on land (bottom left). However, penguins are remarkably swift swimmers. To what nonscientific category do these three types of birds belong?* ❷

112 ■ C

INDEPENDENT PRACTICE
Section Review 4-2

1. Birds are warmblooded, egg-laying vertebrates that have feathers. Feathers are adaptations for flight.
2. Down feathers are short, fluffy feathers that act as insulation. Contour feathers are the largest and most familiar feathers, giving birds their streamlined shape.
3. Birds are similar to reptiles in that both are egg-laying vertebrates that have

CONNECTIONS

Flights of Fancy ❶

When people talk about the beauty of birds, they usually focus on the spectacular colors of the feathers or the lovely melody of the songs. But it is clear that the beauty of the movement of birds has not been lost on the dancers of the world.

For thousands of years, humans have tried to capture the grace and power of birds in *dance*. The mating dance of the crane—with its spectacular leaps, bows, and flapping of wings—is echoed in some of the dances of the Australian aborigines. The majestic, soaring flight of the eagle is re-created in special Native American ceremonial dances. And the graceful, gliding movement of swans while swimming and while in flight is imitated in ballets such as *Swan Lake*.

4–2 Section Review

1. Describe the major characteristics of birds. Which characteristics are adaptations for flight?
2. What is the difference between down and contour feathers? How are feathers adapted to different functions?
3. How are birds similar to reptiles? How are they different?
4. Name and briefly describe four nonscientific categories of birds.

Critical Thinking—*Making Inferences*

5. The kiwi is a chicken-sized bird that is covered with long brownish hairlike feathers and has no visible wings. It has a very long beak, rather short legs, and feet similar to a chicken's. What do you think this bird eats? What can you infer about the kiwi's behavior, environment, and evolution?

ACTIVITY THINKING

A Flock of Phrases

Have you ever been accused of being bird-brained? What other common phrases can you think of that have to do with birds? Get together with a friend or two and see how many you can identify. Based on what you know about birds, determine whether they are accurate.

lungs. The structure and body function processes are more well-developed in birds than in reptiles.

4. The four nonscientific categories of birds include perching birds, birds of prey, water birds, and flightless birds. Perching birds have feet that are well adapted for perching on branches. Birds of prey have excellent vision and are superb fliers with sharp claws and strong, curved beaks. Water birds have webbed feet and broad, flat beaks. Flightless birds have strong leg muscles and legs well adapted to running quickly on flat surfaces.

5. Students might infer that the long beak indicates that a kiwi eats small, hard-to-reach insects; that it has behavior similar to that of other ground-based birds; its strange feathers and small wings indicate that it cannot fly; its flightlessness suggests that it evolved in the absence of significant predators.

Laboratory Investigation

OWL PELLETS

BEFORE THE LAB

1. Well in advance of the activity, order a sufficient supply of owl pellets for groups of two to four students. Owl pellets are available from biological supply companies.
2. Gather all other necessary materials one day prior to the investigation.

PRE-LAB DISCUSSION

Have students read the complete laboratory procedure.

• **What is the purpose of this laboratory activity?** (To systematically classify the types of animals that an owl eats.)

Explain that owls usually feed on small mammals such as those listed in the table. They will also eat small birds. Owls swallow their prey whole, and the soft parts are digested in the stomach. But undigestible parts such as hair, teeth, and bones are regurgitated daily as pellets. An examination of an owl's pellet will provide clues to its feeding habits.

• **As you examine the owl pellet, how will you be able to identify the kinds of animals it has eaten?** (By comparing skulls with the data and pictures in the table.

Laboratory Investigation

Owl Pellets

Problem

What does an owl eat?

Materials

owl pellet	magnifying glass
dissecting needle	small metric ruler

Procedure 🔺 📷

1. Observe the outside of an owl pellet and record your observations.
2. Gently break the pellet into two pieces.
3. Using the dissecting needle, separate any undigested bones and fur from the pellet. Remove all fur from any skulls in the pellet.
4. Group similar bones together in a pile. For example, put all skulls in one group. Observe the skulls. Record the length, number, shape, and color of the teeth.
5. Now try to fit together bones from the different piles to form skeletons.

Observations

1. What does an owl pellet look like? What is it made of?
2. What kinds of bones were the most numerous?
3. What kinds of bones seem to have been missing from the prey?

Analysis and Conclusions

1. What animals were eaten by the owl?
2. Which animals appear to be eaten most frequently by the owl?
3. Why do you think owls cough up pellets?
4. What can you infer about the owl's characteristics from the animals it eats?
5. **On Your Own** Design a study that uses owl pellets to answer a question about the feeding habits of owls. For example, your study might determine whether owls feed on different kinds of prey during different parts of the year.

Shrew	Upper jaw has at least 18 teeth. Skull length is 23 mm or less. Teeth are brown.	
House mouse	Upper jaw has 2 biting teeth. Upper jaw extends past lower jaw. Skull length is 22 mm or less.	
Meadow vole	Upper jaw has 2 biting teeth. Upper jaw does not extend past lower jaw. Molar teeth are flat.	
Mole	Upper jaw has at least 18 teeth. Skull length is 23 mm or more.	
Rat	Upper jaw has 2 biting teeth. Upper jaw extends past lower jaw. Skull length is 22 mm or more.	

TEACHING STRATEGIES

1. Have teams follow the directions carefully as they work in the laboratory.
2. If some students seem reluctant to handle owl pellets, tell them that the pellets have been cleaned to eliminate contamination. Do not force squeamish students to handle the pellets. If left alone, such students will often join in after watching their teammates work for a while.
3. Suggest that as the skulls are separated from the pellets, each is placed on a separate square of paper. Data on each skull can then be written on the same piece of paper.
4. If students are able to fit the bones together to form a skeleton, have them glue these skeletons to a piece of heavy cardboard for display.
5. Remind all students to wash their hands after handling the fur and bones.

DISCOVERY STRATEGIES

Discuss how the investigation relates to the chapter by asking open questions similar to the following:

Summarizing Key Concepts

4–1 Reptiles

▲ Reptiles are vertebrates that have lungs, scaly skin, and a special type of egg.

▲ All living reptiles are coldblooded.

▲ Reptiles have a number of adaptations that make them well suited for life on land.

▲ Reptiles have a double-loop circulatory system. In most reptiles, the two loops are not completely separated.

▲ Land-dwelling reptiles excrete a concentrated nitrogen-containing waste, which helps them to conserve water.

▲ Reptiles have internal fertilization.

▲ A typical reptile egg has a shell and several membranes that protect the developing embryo.

▲ Most reptiles lay eggs; a few reptiles bear live young.

▲ Some reptiles care for their eggs and young.

▲ One scientific group of reptiles includes lizards and snakes.

▲ One scientific group of reptiles is composed of turtles, which are reptiles whose bodies are enclosed by a two-part shell.

▲ One scientific group of reptiles includes crocodiles and alligators.

4–2 Birds

▲ Birds are warmblooded, egg-laying vertebrates that have feathers.

▲ The bodies of birds are adapted for flight. However, not all species of birds can fly.

▲ Feathers, such as contour feathers and down, have shapes that help them to perform their functions.

▲ Bird beaks and feet show a number of interesting adaptations.

▲ The respiratory system is more advanced in birds than in any other vertebrates. The bird's air sacs enable air to be constantly moved through the lungs in one direction.

▲ Birds have a double-loop circulatory system in which the loops are completely separated.

▲ Almost all birds care for their eggs and young.

▲ Birds have many complex behaviors.

▲ One reason birds sing is to establish and maintain a territory.

▲ Many birds migrate.

▲ Four nonscientific—but useful—categories of birds are songbirds, birds of prey, waterfowl, and flightless birds.

Reviewing Key Terms

Define each term in a complete sentence.

4–2 Birds
feather   down
contour feather   territory
                       migrate

much food? (It is warmblooded, and flying expends a great deal of energy—inferring, relating facts.)

OBSERVATIONS

1. Owl pellets are relatively small clumps of organic material.
2. Many students will find mostly pieces of the skull and jawbone.
3. Answers will likely include the smaller bones that may be crushed and broken down into smaller particles during the digestive process.

ANALYSIS AND CONCLUSIONS

1. Answers are likely to include small rodents such as shrews, mice, voles, moles, and rats.
2. Answers will vary but will likely be small rodents.
3. Pellets contain bone and other material that cannot be digested by owls.
4. Students may suggest that owls are birds of prey with keen eyesight, sharp claws, and strong, curved beaks.
5. Individual designs will vary, depending on the student.

GOING FURTHER: ENRICHMENT

Part 1

Tell students that owls regurgitate approximately two pellets each day. Using the number of skulls they found in their pellet, have students determine the number of animals an owl might eat in a year.

Part 2

Ask the class to review the concept of a food web. Then ask them to construct a food web showing the relationship between an owl and the animals it feeds on. In order to trace the feeding levels back to plants, students may need to perform research concerning the feeding habits of prey animals.

• **What adaptations do reptiles have that might enable them to escape from a pursuing owl?** (Answers will vary. Students might suggest that the senses of reptiles can be relatively well developed and that a reptile might become aware of the owl through sight or hearing before the owl strikes. It might also be suggested that reptiles which have the ability to change the color of their body to match the color of their surroundings might be a lesser target—applying, relating concepts.)

Remind students that owls sometimes eat small birds.

• **What adaptations do birds have that might enable them to escape from a pursuing owl?** (Answers will vary. Students might suggest that birds tend to have excellent eyesight and good hearing and may sense the owl before it strikes. It might also be suggested that a small bird may be able to maneuver more quickly than a larger owl—applying, relating facts.)

• **Why does an owl appear to need so**

Chapter Review

ALTERNATIVE ASSESSMENT

The *Prentice Hall Science* program includes a variety of testing components and methodologies. Aside from the Chapter Review questions, you may opt to use the Chapter Test or the Computer Test Bank Test in your *Test Book* for assessment of important facts and concepts. In addition, Performance-Based Tests are included in your *Test Book*. These Performance-Based Tests are designed to test science process skills, rather than factual content recall. Since they are not content dependent, Performance-Based Tests can be distributed after students complete a chapter or after they complete the entire textbook.

CONTENT REVIEW

Multiple Choice

1. a
2. d
3. b
4. c
5. d
6. c
7. b
8. d

True or False

1. T
2. F, egg yolk
3. F, reptiles
4. T
5. F, Crocodiles
6. F, territory
7. F, smell
8. F, perching birds

Concept Mapping

Row 1: Lungs
Row 2: Scaly skin; Lizards and Snakes, Alligators and Crocodiles, Turtles
Row 3: Coldblooded; Egg

Content Review

Multiple Choice

Choose the letter of the answer that best completes each statement.

1. Which of these is a bird of prey?
   a. eagle
   c. toucan
   b. pigeon
   d. ostrich
2. All living reptiles
   a. lay eggs.
   b. spend their entire lives on land.
   c. undergo metamorphosis.
   d. have lungs.
3. The characteristic that separates birds from all other vertebrates is their
   a. claws.
   c. egg-laying.
   b. feathers.
   d. flight.
4. Down feathers
   a. are shaped like leaves.
   b. include large flight feathers on wings.
   c. act as insulation.
   d. all of these

5. Birds sing to
   a. establish a territory.
   b. attract a mate.
   c. warn of danger.
   d. all of these
6. A four-legged water-dwelling shelled reptile that lays leathery-shelled eggs is classified as a(an)
   a. alligator.
   c. turtle.
   b. crocodile.
   d. lizard.
7. To follow seasonal food supplies, birds
   a. perch.
   c. build a nest.
   b. migrate.
   d. hibernate.
8. All reptile eggs
   a. must be laid in water.
   b. have a leathery shell.
   c. are fertilized externally.
   d. contain protective membranes.

True or False

If the statement is true, write "true." If it is false, change the underlined word or words to make the statement true.

1. Because they can find their way back to the beaches on which they hatched, sea turtles can be said to <u>migrate</u>.
2. The part of a reptile or bird egg that is the egg cell is the <u>egg white</u>.
3. Most <u>birds</u> have a <u>double-loop</u> circulatory system in which the loops are not completely separate.
4. Bird adaptations for flight include <u>hollow bones and egg-laying</u>.
5. <u>Alligators</u> have a narrow triangular snout and a toothy grin.
6. A(An) <u>bower</u> is an area in which an individual animal lives.
7. Snakes use their tongues to <u>sting</u>.
8. Songbirds are more correctly known as <u>birds of prey</u>.

Concept Mapping

Complete the following concept map for Section 4–1. Refer to pages C6–C7 to construct a concept map for the entire chapter.

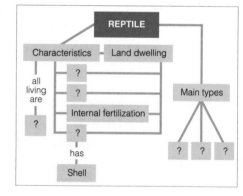

CONCEPT MASTERY

1. Reptile adaptations for land include lungs for breathing, scaly skin to prevent water loss, a water-conserving method of waste elimination, and a protected egg that does not dry out.

2. Sea turtles and many birds migrate regularly to breeding grounds and to follow seasonal food supplies.

3. Reproductive organs in birds are often tiny and compact, reducing weight for flight. Only during the breeding season do the organs enlarge to a functional size. Because birds lay eggs, the female does not carry growing offspring inside her body.

4. A reptile egg typically consists of a hard shell that is sufficiently porous to allow gases to move yet small enough to prevent water from easily escaping, several membranes, egg yolk, and an embryo.

5. The respiratory system of birds exchanges gases with great efficiency because of air sacs attached to their lungs. The circulatory system of birds is a closed double-loop, which ensures that oxygen

Concept Mastery

Discuss each of the following in a brief paragraph.

1. How are reptiles adapted to life on land?
2. Migration may be described as "animals traveling from where they breed to where they feed." How does this description relate to the behavior of sea turtles and many birds?
3. How is reproduction in birds adapted to the demands of flight?
4. What are the major structures of a typical reptile egg?
5. How do the respiratory and circulatory systems of a bird enable it to maintain a highly active lifestyle?
6. Racing pigeons are driven to a place many kilometers from their home and are then released. Even if they have never been at the starting point before, the birds head straight for home. How do the birds find their way back?
7. Parental care in animals ranges from virtually none to quite a lot. Using specific examples, explain how the behavior of reptiles and birds reflects this wide range of behaviors.
8. What is a territory? Why are territories important?

Critical Thinking and Problem Solving

Use the skills you have developed in this chapter to answer each of the following.

1. **Expressing an opinion** Because the skins of alligators, snakes, and some other reptiles make beautiful leather, many species have been hunted to the point of extinction. Do you think the hunting of reptiles should be allowed to continue? Should it be restricted? Should it be stopped altogether?
2. **Developing a hypothesis** Under what circumstances might a live-bearing bird evolve? Explain your reasoning.
3. **Relating concepts** The poisonous coral snake has alternating bands of black, bright red, and bright yellow. The harmless scarlet king snake has a very similar color pattern. Why is this distinctive pattern an advantage to the king snake?
4. **Evaluating theories** Some scientists say that fossil evidence strongly suggests that dinosaurs (long assumed to be coldblooded and an evolutionary dead end) were warmblooded and were the direct ancestors of birds. What sort of information would you need to better evaluate this two-part theory?

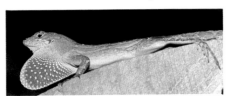

5. **Making inferences** In the spring and summer, male anoles (American chameleons) can be seen throwing their head back and thrusting out a flap of red skin from their throat. What do you think is the purpose of this behavior?
6. **Designing an experiment** A scientist wants to know whether turtles can detect sound. Design an experiment that she can use. Be sure to include a variable and a control in your experiment.
7. **Classifying animals** In the woods, you discover a small four-legged cold-blooded vertebrate. How can you tell whether it is an amphibian or a reptile?
8. **Using the writing process** Imagine that you are the reptile or bird of your choice. Write a short biography describing your life and times.

1. Answers will vary.
2. Hypotheses will vary. A live-bearing bird might evolve if the bird did not need to fly during breeding season and if there were sufficient difficulties with incubation and/or predators that retaining the egg inside the body greatly enhanced the bird's ability to reproduce.
3. Many organisms mistake the king snake for the poisonous coral snake and do not attack it. Thus, by its coloring, the king snake is protected from predators.
4. Answers will vary. Students might suggest that direct evidence of some type would substantiate the theory.
5. It is likely that most students will suggest that it is a courtship behavior.
6. Accept all reasonable designs. Make sure each design includes a variable and a control and that the basic steps of the scientific method are followed.
7. The easiest approach students might suggest is to touch the skin of the vertebrate. If the skin is moist, it is probably an amphibian. If the skin is dry, it is probably a reptile.
8. Biographies will vary, depending on the student. Check each biography for scientific accuracy.

KEEPING A PORTFOLIO

You might want to assign some of the Concept Mastery and Critical Thinking and Problem Solving questions as homework and have students include their responses to unassigned questions in their portfolio. Students should be encouraged to include both the question and the answer in their portfolio.

ISSUES IN SCIENCE

The following issue can be used as a springboard for discussion or given as a writing assignment:

Humans clear away trees and brush for new buildings, highways, and more farmland. This practice destroys the natural habitats of many birds, and the birds may become extinct as a result of this behavior. Should we continue to clear land for buildings, highways, and farms? Why or why not? What is your opinion?

provided by the lungs is delivered effectively to the cells of the body.
6. Answers may vary. Students might suggest that birds such as pigeons navigate by sounds; keen eyesight; using the Earth's magnetic field; using the sun, moon, and stars as guides; and instinct. In all cases, bird navigation is not thoroughly understood by biologists.
7. Examples will vary. Students might suggest that female turtles abandon their eggs after they have been laid and that many songbirds tend their young until the young are old enough to fly and obtain food on their own.
8. A territory is a geographical area that is "claimed" by an organism. This "claim" helps to control the number of inhabitants in an area, causing local food supplies to be sufficient for the organisms allowed within the territory.

Chapter 5　MAMMALS

SECTION	HANDS-ON ACTIVITIES
5–1 What Is a Mammal? pages C120–C123 Multicultural Opportunity 5–1, p. C120 ESL Strategy 5–1, p. C120	**Student Edition** ACTIVITY (Discovering): Vertebrate Body Systems, p. C121 **Teacher Edition** Characteristics of Mammals, p. C118d
5–2 Egg-Laying Mammals pages C123–C124 Multicultural Opportunity 5–2, p. C123 ESL Strategy 5–2, p. C123	
5–3 Pouched Mammals pages C125–C127 Multicultural Opportunity 5–3, p. C125 ESL Strategy 5–3, p. C125	**Student Edition** ACTIVITY (Doing): Endangered Mammals, p. C127
5–4 Placental Mammals pages C127–C137 Multicultural Opportunity 5–4, p. C127 ESL Strategy 5–4, p. C127	**Student Edition** ACTIVITY (Discovering): Migration of Mammals, p. C133 LABORATORY INVESTIGATION: Examining Hair, p. C138 **Teacher Edition** The Opposable Thumb, p. C118d **Laboratory Manual** Identifying Vertebrates Using Classification Keys, p. C101 The Most Intelligent Mammals, p. C107
Chapter Review pages C138–C141	

OUTSIDE TEACHER RESOURCES
Books

Book of Mammals, National Geographic Society.

Burton, John S., and Bruce Pearson. *The Collins Guide to the Rare Mammals,* Stephen Greene Press.

Grzimek, Dr. Bernhard. *Grzimek's Encyclopedia: Mammals,* McGraw-Hill.

MacDonald, David, ed. *The Encyclopedia of Mammals,* Facts On File.

Walker, E. P. *Mammals of the World,* 3rd ed., Johns Hopkins University Press.

Wild Animals of North America, National Geographic Society.

Zappler, Lisbeth, and Georg Zappler. *The World After the Dinosaurs: The Evolution of Mammals,* The Natural History Press.

OTHER ACTIVITIES	MEDIA AND TECHNOLOGY
Student Edition ACTIVITY (Reading): A Cat and Two Dogs, p. C123 **Activity Book** CHAPTER DISCOVERY: Examining Teeth, p. C95 **Review and Reinforcement Guide** Section 5–1, p. C63	**Video/Videodisc** Sound Sense **English/Spanish Audiotapes** Section 5–1
Review and Reinforcement Guide Section 5–2, p. C65	**Courseware** Classifying Mammals (Supplemental) **English/Spanish Audiotapes** Section 5–2
Review and Reinforcement Guide Section 5–3, p. C67	**Video** Mammals (Supplemental) **English/Spanish Audiotapes** Section 5–3
Student Edition ACTIVITY (Writing): Useful Mammals, p. C129 ACTIVITY (Calculating): The Fastest Runner, p. C130 **Activity Book** ACTIVITY: Find the Oddball, p. C99 ACTIVITY: Scoring Vertebrates, p. C103 ACTIVITY: Classifying Animals, p. C107 ACTIVITY: Classifying Mammals, p. C109 **Review and Reinforcement Guide** Section 5–4, p. C69	**Interactive Videodisc** In the Company of Whales ScienceVision: EcoVision **Video/Videodisc** The Mystery of a Million Seals (Supplemental) **Video** The Great Bears (Supplemental) The Passage of Migrants (Supplemental) **Courseware** Food Chains and Webs (Supplemental) **English/Spanish Audiotapes** Section 5–4
Test Book Chapter Test, p. C91 Performance-Based Tests, p. C109	**Test Book** Computer Test Bank Test, p. C97

*All materials in the Chapter Planning Guide Grid are available as part of the Prentice Hall Science Learning System.

Audiovisuals

Australia's Unusual Animals, film or video, National Geographic

Egg-Laying Mammals: The Echidnas and Platypus, film or video, AIMS Media

Introduction to the Mammals, film or video, Carolina Biological Supply Co.

Mammals, film or video, AIMS Media

Mammals, film or video, Coronet

Mammals, Life Cycles series, filmstrip with cassette, National Geographic

Mammals, Part II: Vertebrates, filmstrip with cassette, National Geographic

What Is a Mammal? film or video, EBE

Chapter 5 MAMMALS

CHAPTER OVERVIEW

The group of animals known as mammals includes organisms that vary greatly in size and habits. But they all have certain characteristics in common. Mammals are warmblooded vertebrates that have hair or fur at some time in their life and that feed their young with milk produced in mammary glands. This last characteristic gives mammals their name. Mammals have lungs and highly developed hearts, excretory systems, brains, and sense organs.

Over time, mammals evolved into three distinct groups: monotremes, marsupials, and placental mammals. Reproduction varies among the three groups. Monotremes are egg-layers. Duckbill platypuses and spiny anteaters are monotremes. Marsupial young are born at a very early stage of development and must develop further in a pouch. Kangaroos, koalas, and opossums are marsupials. Placental young develop more fully within the body of the mother. There are more kinds of placental mammals than any other mammals. Ten groups of placental mammals are described in this text: insect-eating (mole, hedgehog, shrew); flying (bat); flesh-eating (walrus, seal); toothless (anteater, armadillo, sloth); trunk-nosed (elephant); hoofed (pig, horse); gnawing (rat, mouse); rodentlike (rabbit, hare, pika); water-dwelling (whale, manatee); and primates (chimpanzee, human).

Mammals evolved from early reptile ancestors. They are found today throughout the world in almost every kind of environment.

5-1 WHAT IS A MAMMAL?

THEMATIC FOCUS

The purpose of this section is to introduce students to the group of animals called mammals. There are about 4000 different kinds of mammals on Earth. Although they are all very different, they can be grouped together because they share some important characteristics. Students will learn that mammals are warmblooded, they are vertebrates, they have hair or fur, and they feed their young with milk produced in mammary glands. They provide their young with more care than do other animals. Mammals use lungs to breathe; they have a four-chambered heart, a well-developed excretory system, and a brain that is the most highly developed of all the animals. Mammals also have highly developed senses of sight, hearing, taste, and smell. Students will discover that the three main groups of mammals differ in the way they reproduce: egg-laying mammals lay eggs, pouched mammals give birth to young that are not well developed and must live in a pouch for a time, and placental mammals give birth to young that have fully developed inside their mother.

The themes that can be focused on in this section are energy, unity and diversity, systems and interactions, and stability.

Energy: Stress that all mammals obtain energy by eating plants and/or other animals.

*Unity and Diversity: It is important to remember that the differences in the way mammals reproduce provide a means of classifying them into three groups: monotremes, marsupials, and placental mammals.

Systems and interactions: Point out that mammals respond to and interact with their environments in ways that help them to gather food, reproduce, and protect themselves. Over time, they have acquired features and behavior that enable them to survive in their environments.

Stability: The various life functions of mammals serve to maintain a stable internal and external environment; that is, everything from how mammals breathe, move, and think to what they eat and where they live is geared toward survival.

PERFORMANCE OBJECTIVES 5-1

1. Describe the main characteristics of mammals.
2. Explain why mammals are thought to be the most intelligent animals on Earth.
3. List the three basic groups of mammals.

SCIENCE TERMS 5-1

egg-laying mammal p. C123
pouched mammal p. C123
placental mammal p. C123

5-2 EGG-LAYING MAMMALS

THEMATIC FOCUS

The purpose of this section is to introduce students to egg-laying mammals, or monotremes. The only known monotremes are the duckbill platypus and the spiny anteater. Even though monotremes lay eggs as reptiles do, they do not leave the hatchlings to fend for themselves as reptiles do. Young platypuses and spiny anteaters feed on their mothers' milk as all young mammals do. Students will learn about the unusual physical characteristics of these remarkable and rare creatures.

The themes that can be focused on in this section are evolution and unity and diversity.

*Evolution: The first mammals, which appeared 200 million years ago, evolved from a now-extinct group of reptiles. Exactly how monotremes, marsupials, and placental mammals evolved from those early mammals is still unknown. Monotremes, or egg-laying mammals, are regarded as the most primitive mammals.

*Unity and diversity: Monotremes may be unusual mammals because they lay eggs as birds and reptiles do. Nevertheless, they are mammals because they feed their young on milk from mammary glands.

PERFORMANCE OBJECTIVES 5-2

1. Name the two kinds of egg-laying mammals.

2. List the major traits of the egg-laying mammals.

3. Compare the feeding habits of the spiny anteater and the platypus.

5-3 POUCHED MAMMALS
THEMATIC FOCUS

The purpose of this section is to introduce students to pouched mammals, or marsupials. Marsupials give birth to young that are not well developed. The young must stay in a pouchlike structure in their mother's body until they have developed completely. Students will learn specifically about koalas, kangaroos, and opossums. Koalas and kangaroos are marsupials native only to Australia. Opossums are the only marsupial in North America.

The theme that can be focused on in this section is unity and diversity.

Unity and diversity: Pouched animals, or marsupials, are different from other mammals because they give birth to young that are not well developed. These babies must find their way to the mother's pouch, where they remain until they are completely developed. Nevertheless, marsupials are mammals because they feed their young on milk from mammary glands.

PERFORMANCE OBJECTIVES 5-3

1. Identify the major characteristic of marsupials.

2. Describe three familiar marsupials.

5-4 PLACENTAL MAMMALS
THEMATIC FOCUS

The purpose of this section is to introduce students to the third and largest group of mammals: placental mammals. These animals give birth to young that have developed completely while inside their mother. A structure called a placenta allows food, water, and wastes to be exchanged between the young and the mother. The gestation period, or the time the young spend inside the mother, varies from mammal to mammal. After the young are born, they are fed milk from their mother's mammary glands, just as monotreme and marsupial young are. Students will learn about ten groups of placental mammals: insect-eating, flying, flesh-eating, toothless, trunk-nosed, hoofed, gnawing, rodentlike, water-dwelling, and primates. These mammals are organized according to how they eat, how they move about, or where they live. But despite their many differences, they are all mammals.

The theme that can be focused on in this section is unity and diversity.

Unity and diversity: Placental mammals differ from monotremes and marsupials in the way they reproduce. Placental mammals also differ greatly from one another; there are 16 different groups of mammals, ranging from bats to whales to gorillas. Nevertheless, they are all mammals because they give birth to young who have developed fully functioning body systems while inside the mother and because they feed their young on milk produced by mammary glands.

PERFORMANCE OBJECTIVES 5-4

1. Describe the characteristics of placental mammals.

2. Classify ten groups of placental mammals and give an example of each.

3. Compare placental mammals to monotremes and marsupials.

4. Compare carnivorous and herbivorous placental mammals.

SCIENCE TERMS 5-4

placenta p. C127
gestation period p. C128

Discovery *Learning*

TEACHER DEMONSTRATIONS MODELING

Characteristics of Mammals

Obtain a live mammal such as a hamster, a guinea pig, or a mouse. Show the animal to the class. Allow one or two students to touch or hold the animal gently.

• **What do you notice when you touch this animal?** (It is warm and has fur.)

• **Warmblooded animals that have hair are mammals. How many of you have pet mammals? What kind of mammal do you have?** (Answers will vary. You may wish to write the names of the different mammals on the chalkboard.)

• **Has your pet ever had babies? How did your pet take care of its babies?** (Answers will vary. Lead students to note that their pet nursed its young.)

Tell the class that mammals, as well as having hair and being warmblooded, also feed their young with milk from mammary glands.

• **What are some other animals that are mammals?** (Write all appropriate mammals on the chalkboard.)

• **How are these animals alike?** (They have hair, are warmblooded, and feed their young with milk. Other answers are also appropriate.)

• **How are these animals different?** (Accept all reasonable answers.)

Tell students that they will learn more about the traits of mammals in this chapter.

The Opposable Thumb

Show students the palms of your hands with your fingers and thumbs extended. Fold your thumb into the palm and back out several times. Touch each of your fingers in turn with the thumb of that hand. Form your hand into a fist and open it several times. Pick up an object.

• **Can all of you do these exercises?** (Yes.)

• **Although these exercises are easy for us, they are unique in the world of animals. Why do you think they are unique?** (Accept all answers.)

• **How does being able to move a thumb make us different from most other living things?** (Accept all answers.)

CHAPTER 5
Mammals

INTEGRATING SCIENCE

This life science chapter provides you with numerous opportunities to integrate other areas of science, as well as other disciplines, into your curriculum. Blue numbered annotations on the student page and integration notes on the teacher wraparound pages alert you to areas of possible integration.

In this chapter you can integrate life science and evolution (pp. 120, 122), language arts (pp. 123, 129, 132), social studies (p. 126), life science and classification (p. 128), physical science and sonar (p. 129), mathematics (p. 130), fine arts (p. 133), life science and disease (p. 134), and physical science and sound (p. 137).

SCIENCE, TECHNOLOGY, AND SOCIETY/COOPERATIVE LEARNING

Mammals are one of the largest groups of animals that are hunted by humans. There are three types of hunting: commercial hunting, hunting for food, and recreational hunting. Commercial hunting is used to supply large quantities of a particular type of animal for the public. Unregulated and irresponsible commercial hunting can lead to the endangerment or extinction of some species; the American bison is a good example. Hunting for food, once a way of life for people in the United States, is still a necessity in many parts of the world today.

In the United States, most individuals hunt for sport or recreation. Advocates of recreational hunting say that they enjoy the challenge and sportsmanship, learn about wildlife, and gain greater respect for nature. Taxes and license fees from hunters also supply some of the money used by states for wildlife management. Hunters also argue that hunting maintains the balance between populations in the environment and that careful control by state agencies prevents overhunting of any one species.

Opponents of hunting for sport argue that it is a cruel and unnecessary practice that endangers wildlife and their habitats. They also cite the dangers involved in the use of high-powered rifles.

INTRODUCING CHAPTER 5

DISCOVERY LEARNING

▶ *Activity Book*

Begin teaching the chapter by using the Chapter 5 Discovery Activity from the *Activity Book*. Using this activity, students will discover that teeth can be used to determine an animal's diet and thus to classify animals.

USING THE TEXTBOOK

Have students look at the photograph on page C118.
• **Do you recognize this animal? What is it doing?** (Students may recognize the animal as a sea otter. It is floating on its back amid numerous strands of giant seaweed.)
• **Does the otter remind you of any other animals you know?** (Accept all logical answers. Students may mention a resem-

Mammals

Guide for Reading

After you read the following sections, you will be able to

5–1 What Is a Mammal?
■ Describe the main characteristics of mammals.

5–2 Egg-Laying Mammals
■ Identify the characteristics of egg-laying mammals.

5–3 Pouched Mammals
■ Describe the characteristics of pouched mammals.
■ Compare egg-laying and pouched mammals.

5–4 Placental Mammals
■ Describe the characteristics of placental mammals.
■ Classify ten groups of placental mammals and give an example of each.

A few hundred meters off the coast of California, a small group of animals swim playfully with one another. These whiskered animals are sea otters. A sea otter spends most of its life swimming in the cold waters of the North Pacific Ocean. While floating on its back, a sea otter often balances a rock on its chest. It is against this rock that the sea otter strikes a closed clam shell, cracking open the shell and eating the clam inside.

A sea otter appears to be an intelligent animal. To keep from being swept away by waves, a sea otter wraps itself in strands of giant seaweed growing offshore. It uses the seaweed as giant ropes are used to hold an ocean liner close to a pier.

Although sea otters spend a great deal of time in the water, they are neither fishes nor amphibians. Sea otters belong to the same group of warmblooded vertebrates that you do: the mammals. In addition to swimming in the sea, mammals can be found flying in the air and running along the ground. To learn more about these remarkable creatures, just turn the page.

Journal *Activity*

You and Your World If you have a pet mammal such as a dog, cat, hamster, rat, mouse, gerbil, horse, or guinea pig, observe it for a day. If you do not have a pet, observe one of your friend's pets. In your journal, record all the animal's activities and the time it spends doing each activity. Include a photograph of the animal in your journal.

◀ *A sea otter floating on strands of giant seaweed*

Cooperative learning: Using preassigned lab groups or randomly selected teams, have groups complete one of the following assignments.

• Have groups find out which mammals are legally hunted in their state and the restrictions placed on hunters of each kind of mammal. What changes, if any, would they recommend to the state agency in charge of hunting regulations?
• Post the following reaction statement in the classroom: Wildlife is a public resource, and the public has the right to harvest it. After the groups discuss the statement, have them share their reaction to the statement in one of the following ways: T-shirt legend, billboard, letter to the editor or congressional representative, bumper sticker, and the like. If time permits, you might want to assign each group a position on this issue and stage a classroom debate.

See Cooperative Learning in the *Teacher's Desk Reference.*

JOURNAL ACTIVITY

You may want to use the Journal Activity as the basis of a class discussion. Encourage students to compare the activities of the mammals they observed. Students who observed the same kind of mammal can work together to compile a list of common activities. All students can make a chart that lists the activities by mammal and indicates the similarities in behavior. Students should be instructed to keep their Journal Activity in their portfolio.

---

blance to squirrels, beavers, seals, mice, and so on.)

List the animals on the chalkboard as students name them.
• **What do these animals, including the otter, have in common?** (Accept all logical answers. Students may concentrate on features such as nose, whiskers, paws, and so forth. Lead them to note that all the animals have fur.)

Have students read the chapter introduction on page C119. Point to the list of animals on the chalkboard.
• **Now what else do you know that these animals have in common?** (They are warmblooded; they are vertebrates.)
• **What does *vertebrates* mean?** (They have backbones.)
• **Otters live mostly in the sea, whereas squirrels live entirely on land. How can**

they both be mammals? (According to the introduction, mammals can live in many different places. Some live in the water, some live on land, and some can even fly.)

Point out to students that although mammals have some things in common, they are also very different animals. Students will learn more about these animals in this chapter.

5-1 What Is a Mammal?

The distribution of mammals around the world is an interesting study. One might think that all tropical areas would have the same kinds of mammals, but this is not the case. The tropical areas of America are the only place in the world where there are sloths and monkeys with prehensile tails. In these same regions, however, there are no apes, elephants, or rhinos.

Have students investigate the animal life of Australia and New Zealand and discuss why some of the world's most interesting animals are found there. Scientists who study evolutionary patterns have suggested that a number of animals in this region evolved uniquely because the region was cut off from other contiguous land masses in early times.

ESL STRATEGY 5-1

Suggest that students make a comparison chart of vertebrate characteristics. Have them work in pairs to complete the chart. When discussing mammals' teeth, make sure that students can differentiate between incisors and canines and their uses. Explain the difference between bite or grasp and tear or shred.

TEACHING STRATEGY 5-1

FOCUS/MOTIVATION

Have students rub one of their hands up and down one of their arms.
• **What do you feel?** (Accept all logical answers. Lead students to realize that they can feel the hair on their arms.)

Divide students into groups of two or three. Have students look at the hair on their arms without a magnifying glass.
• **What do you see?** (Accept all logical answers.)

Give each group a magnifying glass. Have students observe the hair and pores of their arms through the magnifying glass.
• **What does it look like now?** (Accept all logical answers.)
• **Does the amount of hair vary from person to person?** (Yes.)
• **Do you think humans are unique be-**

Guide for Reading

Focus on this question as you read.
▶ What are the main characteristics of mammals?

Figure 5-1 The first mammals to appear on Earth may have resembled the modern-day tree shrew (top). What characteristics common to all mammals are illustrated by the moose (bottom right) and the musk ox (bottom left)? ❶

cause of the hair that is on their bodies? (Accept all logical answers.)

CONTENT DEVELOPMENT

Point out that humans as well as lions, bears, bats, and seals all belong to a class of chordates called mammals. Explain that mammals are warmblooded vertebrates that have hair and feed their young with milk produced in mammary glands. Tell students that the mammary gland is a special gland in a mammal that produces milk.

5-1 What Is a Mammal?

About 200 million years ago, the first mammals appeared on Earth. They evolved from a now-extinct group of reptiles. The first mammals were very small and looked something like the modern-day tree shrew shown in Figure 5-1.

Today there are about 4000 different kinds of mammals living on Earth. In addition to humans and sea otters, mammals include whales, bats, elephants, duckbill platypuses, lions, dogs, kangaroos, and monkeys. Because scientists group together animals with similar characteristics, you might wonder what such different-looking animals have in common.

Mammals have characteristics that set them apart from all other living things. **Mammals are warm-blooded vertebrates that have hair or fur and that feed their young with milk produced in mammary glands.** In fact, the word mammal comes from the term mammary gland. Another special characteristic of mammals is that they provide their young with more care and protection than do other animals.

At one time during their lives, all mammals possess fur or hair. If it is thick enough, the fur or hair acts as insulation and enables mammals such as musk oxen to survive in very cold parts of the world. Musk oxen are the furriest animals alive today. Indeed, the fur of an adult musk ox may be as deep as 15 centimeters! Mammals can also survive in harsh climates because they are warmblooded. Recall from Chapter 3 that warmblooded animals maintain their body temperatures internally as a result of the chemical reactions that occur within their cells. Thus mammals

Point out that the structure of hair changes on some mammals. Explain that the wool of sheep, the quills of porcupines, and the armor of armadillos are all modified types of hair.

● ● ● ● **Integration** ● ● ● ●

Use the discussion of the appearance of the first mammals on Earth to integrate concepts of evolution into your lesson.

maintain a constant body temperature despite the temperature of their surroundings. What other group of animals can do this? ❷

All mammals, even those that live in the ocean, use their lungs to breathe. The lungs are powered by muscles—a group of muscles that are attached to the ribs and one large muscle that separates the abdomen from the chest.

The circulatory system of mammals consists of a four-chambered heart and an assortment of blood vessels. The heart pumps oxygen-poor blood to the lungs, where the blood exchanges its carbon dioxide for oxygen. After leaving the lungs, the oxygen-rich blood returns to the heart and is pumped to all parts of the body through blood vessels.

Mammals have the most highly developed excretory system of all the vertebrates. Paired kidneys filter nitrogen-containing wastes from the blood in the form of a substance called urea (yoo-REE-uh). Urea combines with water and other wastes to form urine. From the kidneys, urine travels to a urinary bladder, where it is stored until it passes out of the body.

The nervous system of mammals consists of a brain that is the most highly developed of all the animals. The brain makes thinking, learning, and understanding possible; coordinates movement; and regulates body functions. Mammals also have highly

Figure 5–2 *The large, flat, grinding teeth of a white-tailed deer indicate that this mammal eats plants. The sharp, pointed teeth of a gray wolf indicate that this mammal eats the flesh of its prey.*

ACTIVITY
DISCOVERING

Vertebrate Body Systems

Vertebrates have well-developed body systems that show considerable diversity from one group to another. Choose one of the systems listed below and illustrate how it changes from fishes to amphibians to reptiles to birds to mammals.

Systems: nervous system, digestive system, circulatory system, reproductive system, excretory system

■ Why do you think vertebrate systems become more complex?

C ■ 121

ACTIVITY
READING

A CAT AND TWO DOGS

Skill: Reading comprehension

You may want to have a reading round-table discussion of the book by those students who read the book.

Integration: Use this Activity to integrate language arts skills into your science lesson.

developed senses that provide them with information about their environment. For example, humans, monkeys, gorillas, and chimpanzees are able to see objects in color. This characteristic is extremely useful because these mammals are most active during the day when their surroundings are bathed in light. Many mammals—cats, dogs, bats, and elephants, for example—are more sensitive to certain sounds than humans are.

Mammals also have more highly developed senses of taste and smell. For example, humans use both their sense of taste and their sense of smell to determine the flavor of food. Dogs and cats, as you might already know, recognize people by identifying specific body odors.

Like reptiles and birds, all mammals have internal fertilization, and males and females are separate individuals. However, the way in which mammals reproduce differs. The differences in reproduction

Figure 5-3 *The brain of a mammal is large compared to that of other animals.*

Figure 5-4 *The 4500 species of mammals can be divided into three main groups. What are the names of the three main groups?* ❶

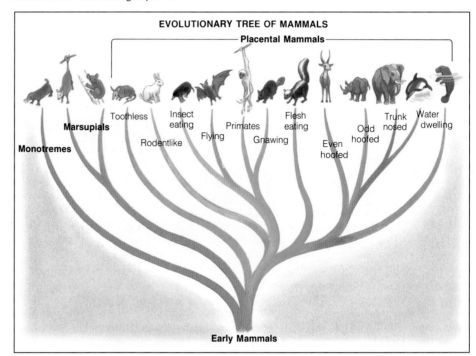

EVOLUTIONARY TREE OF MAMMALS

Placental Mammals

Marsupials

Monotremes

Toothless

Rodentlike

Insect eating

Flying

Primates

Gnawing

Flesh eating

Even hoofed

Odd hoofed

Trunk nosed

Water dwelling

Early Mammals

122 ■ C

5-1 (continued)

CONTENT DEVELOPMENT

Ask students to look at Figure 5–4. Explain that this diagram shows how mammals are divided into different groups. Point to the three main branches leading to the egg-laying mammals, the pouched mammals, and the placental mammals. Explain that placental mammals are divided into a number of groups, which students will learn about later in this chapter.

● ● ● ● **Integration** ● ● ● ●

Use the discussion of mammal groups to integrate concepts of evolution into your lesson.

INDEPENDENT PRACTICE

Section Review 5-1

1. They are warmblooded vertebrates with hair or fur who feed their young with milk produced in mammary glands.

2. They produce milk to feed the young mammals.

3. The three groups of mammals are classified according to the way they reproduce. Egg-laying mammals lay eggs. Pouched mammals give birth to young who must complete their development in a pouch in their mother's body. Placental mammals give birth to young who have developed their own body systems while inside the mother.

4. Maintaining a constant body temperature during cold weather requires that the mammal's body use large amounts of

provide a means of classifying mammals into three main groups. These groups are **egg-laying mammals, pouched mammals,** and **placental** (pluh-SEHN-tuhl) **mammals.** Egg-laying mammals, as their name implies, lay eggs. Pouched mammals give birth to young that are not well developed. Thus the young must spend time in a pouchlike structure in their mother's body. In placental mammals, the young remain inside the mother until their body systems are able to maintain life on their own. At birth, these young are more developed than are those who spend time in their mother's pouch. You will learn more about each group of mammals in the remainder of this chapter.

5–1 Section Review

1. What are the main characteristics of mammals?
2. What is the function of mammary glands?
3. Classify the three groups of mammals.

Critical Thinking—*Relating Concepts*
4. Why does a mammal need more food during cold weather than during warmer weather?

5–2 Egg-Laying Mammals

One of the strangest looking mammals on the Earth today lives in rivers in isolated parts of Australia. It has fur as thick as a sea otter's, feet that are weblike and clawed, and a large flat ducklike beak! What is this strange creature?

If you look at Figure 5–5 on page 124, you will discover what this weird-looking animal is. It is a duckbill platypus. What makes the duckbill platypus stranger still is that although it is a mammal, it lays eggs! **Mammals that lay eggs are called egg-laying mammals, or monotremes** (MAHN-oh-treemz). Because of their ability to lay eggs, egg-laying mammals are sometimes referred to as reptilelike or primitive. The duckbill platypus and the spiny anteaters are the only known monotremes.

A Cat and Two Dogs

Sheila Burford is the author of a wonderful book about a cat and two dogs that make their way home through the wilds of Canada. The book is entitled *The Incredible Journey.*

Guide for Reading

Focus on this question as you read.

▶ *What are egg-laying mammals?*

C ■ 123

5-2 Egg-Laying Mammals

MULTICULTURAL OPPORTUNITY 5-2

Animals play prominent roles in folk tales, myths, and legends of many cultures including the Vietnamese. Mai-Vo-Dinh, a writer known for his tales about animals, was born in Hue, Vietnam, in 1933 and came to the United States in 1960, where he eventually became an American citizen. As a child in Vietnam, he knew many of his country's legends and folk tales. As an adult, he remembers them vividly and retells them in his interesting collection, *The Toad Is the Emperor's Uncle: Animal Folk Tales From Vietnam.* Vo-Dinh is also a book illustrator whose paintings and woodcuts, which frequently include animals, have been exhibited in the United States and abroad. Suggest that students acquaint themselves with the work of this writer and artist. Interested students may also want to find out more about Vietnam and the importance of nature in Vietnamese literature and folklore.

ESL STRATEGY 5-2

Have students write a short paragraph comparing how a duckbill platypus's and a spiny anteater's eggs are hatched. Ask an English-speaking student to act as tutor in helping an ESL student write the paragraph.

• **What kinds of animals do these remind you of?** (Accept all reasonable answers.)
• **How are these animals similar to other mammals?** (They have hair. Lead students to infer that the monotremes also share other mammalian traits.)
• **What sort of things can you tell about these animals just by looking at them?** (Accept all reasonable answers. Students might guess that spiny anteaters are protected by their spines, that platypuses swim because they have webbed feet, and that spiny anteaters can eat only small things because their mouths are small.)

energy produced from the food that the mammal eats. The more energy required, the more food the mammal will need to eat.

REINFORCEMENT/RETEACHING

Monitor students' responses to the Section Review questions. If students appear to have difficulty understanding any of the concepts, review this material with them.

CLOSURE

▶ *Review and Reinforcement Guide*

At this point have students complete Section 5–1 in the *Review and Reinforcement Guide.*

TEACHING STRATEGY 5-2

FOCUS/MOTIVATION

Obtain photographs and/or pictures of monotremes. Have students examine these photographs and pictures.

HISTORICAL NOTE

DUCKBILL WHAT?

The platypus was first seen by European settlers in Australia in 1797. When the first specimen, a dried skin, was sent to the Natural History Museum in London in 1798, scientists thought it was a hoax, a fake made of different animal bits stitched together! Even when the platypus was found to be real, scientists did not accept it as a mammal—after all, it laid eggs—until they discovered that the platypus had mammary glands, the major distinguishing feature of mammals.

Figure 5-5 *The duckbill platypus (top) and the spiny anteater (bottom) are two of the six species of egg-laying mammals that exist today. Egg-laying mammals live in isolated parts of Australia and New Guinea. What is another name for egg-laying mammals?* ①

When a female duckbill platypus lays her soft marble-sized eggs, which usually number from one to three, she deposits them in a burrow she has dug in the side of a stream bank. The female will keep the eggs warm for the 10 days it takes them to hatch. Once hatched, the young platypuses are not left to find food for themselves (as are the young of reptiles). Instead, the young platypuses feed on milk produced by their mother's mammary glands, which are located on her abdomen. Milk production, as you may recall, is a characteristic of mammals.

Soon after a female spiny anteater lays her eggs, she places them into a pouch on her abdomen. The eggs hatch in 7 to 10 days. Like young platypuses, young spiny anteaters feed on milk produced by their mother's mammary glands.

The unusual body parts of a duckbill platypus help it to gather its food. For example, a duckbill platypus uses its claws to dig for insects, then uses its soft ducklike bill to scoop them up. The bill serves another important purpose: When under water, a platypus closes its eyes and ears. Unable to see as it swims above the riverbed, the platypus feels for snails, mussels, worms, and sometimes small fishes with its bill, which is very sensitive to touch.

Spiny anteaters also have special structures that help them to gather their food: ants and termites. A spiny anteater has a long, thin snout that it uses to probe for food, and it has a sticky, wormlike tongue that it flips out to catch insects. To protect itself, a spiny anteater uses its short powerful legs and curved claws to dig a hole in the ground and cover itself until only its spines are showing. These spines, which are 6 centimeters long, usually discourage almost any enemy.

5-2 Section Review

1. What is a characteristic of egg-laying mammals?
2. Name the two types of egg-laying mammals.

Critical Thinking—*Applying Concepts*

3. Why might egg-laying mammals be considered a link between reptiles and mammals?

5-2 (continued)

CONTENT DEVELOPMENT

Monotremes and certain shrews are the only venomous mammals. In the spiny anteater, the structure that produces and delivers the venom is vestigial. Male platypuses have a bony spur on the back of each ankle. This spur is connected to a venom-producing gland. A jab from the spur can kill a dog and can cause severe pain in a human.

INDEPENDENT PRACTICE

Section Review 5-2

1. Unlike other mammals, they lay eggs.
2. The duckbill platypus and the spiny anteater.

3. Because they lay eggs as reptiles do and they feed their young milk produced in mammary glands as mammals do.

REINFORCEMENT/RETEACHING

Review students' responses to the Section Review questions. Reteach any material that is still unclear, based on their responses.

CLOSURE

▶ *Review and Reinforcement Guide*

Students may now complete Section 5-2 in the *Review and Reinforcement Guide.*

TEACHING STRATEGY 5-3

FOCUS/MOTIVATION

Tell students that an adult common opossum is about the same size as a cat.

5–3 Pouched Mammals

Unlike egg-laying mammals, pouched mammals do not lay eggs. Instead, they give birth to young that are not well developed. Thus the young must spend time in a pouchlike structure in their mother's body. Mammals that have pouches are called **marsupials** (mahr-soo-pee-uhlz).

When most people hear the word marsupial, they think of a kangaroo. Kangaroos, however, are not the only pouched mammals. Pouched mammals also include koalas, opossums, wombats, bandicoots, and gliders. Figure 5–6 shows some of these pouched mammals.

Perhaps the cuddliest and cutest pouched mammal is the koala. See Figure 5–7 on page 126. The koala's ears are big and round and covered with thick fur. Unlike many pouched mammals, a koala has a pouch whose opening faces its hind legs rather than its head. Koalas spend most of their time in trees, munching away on the only food they eat—the

Guide for Reading

Focus on these questions as you read.

▶ What are pouched mammals?

▶ What are the similarities and differences between egg-laying mammals and pouched mammals?

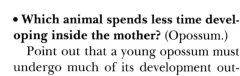
Figure 5–6 *The hairy-nosed wombat (bottom right), the eastern barred bandicoot (top left), and the sugar glider (bottom left) are examples of pouched mammals.*

5–3 Pouched Mammals

MULTICULTURAL OPPORTUNITY 5-3

Animals, fishes, birds, and reptiles are important in various aspects of Japanese culture. Students will be interested to learn that in Japan, each year in a 12-year cycle bears the name of a real or an imaginary animal. The Japanese Year Names are as follows: 1—rat, 2—ox, 3—tiger, 4—rabbit, 5—dragon, 6—snake, 7—horse, 8—sheep, 9—monkey, 10—cock, 11—dog, 12—wild boar. The same animal symbols are used to identify the hours of the day; for example, 2 AM is the Ox Hour and 11 PM is the Dog Hour. These animals also appear in Japanese folklore, literature, and art. The rat, for instance, is frequently depicted in Japanese artistic metalwork and in paintings; the ox figures prominently in tales and proverbs; the wild boar is often mentioned in history and literature. Suggest that students find out about the symbolic significance of each animal represented by the Japanese Year Names.

ESL STRATEGY 5-3

You may want to write the following sentences on the chalkboard and ask for volunteers to complete them. Have students read the completed sentences aloud.

1. A monotreme is a(n) ____-_____ mammal.

2. A _____ is a pouched mammal that eats plants; therefore, it is known as a(an) _____.

A newborn opossum, however, is about 1 centimeter long, whereas a newborn kitten is about 7 centimeters long. If possible, obtain photographs of newborn kittens and newborn opossums to show the class.

- **Which animal is larger at birth?** (Cat.)
- **Which animal would you expect to be less developed at birth? Why?** (Opossum, because it's much smaller.)
- **Where do mammals develop before birth?** (Inside the mother.)

- **Which animal spends less time developing inside the mother?** (Opossum.)

Point out that a young opossum must undergo much of its development outside the mother's body.

- **How might the opossum care for its newborn offspring?** (Accept all reasonable answers. Lead students to suggest that an opossum needs a place where the young can be kept warm and protected while they develop.)

CONTENT DEVELOPMENT

Point out that animals with pouches are called marsupials. Explain that marsupials are born only partially developed. After birth, the newborn babies immediately move into a special body pouch near the mother's nippled mammary glands. Tell students that marsupial babies feed on the milk from the mother's mammary glands until they are developed enough to leave the pouch.

ACTIVITY
DOING
ENDANGERED MAMMALS

Skills: Applying concepts

Materials: reference materials

Endangered species are living things that are threatened with extinction. Most biologists consider a species endangered if they expect it to die off completely in less than 20 years without special efforts to protect it. Endangered mammals include Asiatic lions, black-footed ferrets, black rhinoceroses, blue whales, orangutans, red wolves, snow leopards, and tigers. Students' suggestions should be logical.

leaves of eucalyptus trees. If you have ever sucked cough drops, you are probably familiar with the scent of eucalyptus leaves. The oils of the eucalyptus leaves are put into cough drops. So it is no wonder that koalas smell like cough drops! Because koalas eat only plant material, they, like most marsupials, are called herbivores (HER-buh-vorz).

Kangaroos

"At best it resembles a jumping mouse but it is much larger." These words were spoken in 1770 by the English explorer James Cook. He was describing ① an animal never before seen by Europeans. He later gave the name "kangaroo" to this strange animal.

Kangaroos live in the forests and grasslands of Australia. They have short front legs but long, muscular hind legs and tails. The tail helps the kangaroo to keep its balance and to push itself forward.

When a kangaroo is born, it is only 2 centimeters long and cannot hear or see. Although only partially developed at birth, it has front legs that enable it to crawl as many as 30 centimeters to its mother's pouch. To get a better idea of just what this feat involves, imagine the following: You are blindfolded, your ears are plugged, and you are placed in the center of a strange room about the size of half a football field. You are then told to find the exit on the first try! Obviously, this task is not easy. In fact, it is almost impossible. But the young kangaroo does it—and successfully, too. The young kangaroo, which is called a joey, stays in its mother's pouch for about 9 months, feeding on its mother's milk.

Figure 5–7 *Pouched mammals, such as the koala, the kangaroo, and the opossum, give birth to young that are not well developed. Thus the young must spend time in a pouchlike structure in their mother's body. Notice the baby kangaroo, called a joey, in its mother's pouch. What is another name for pouched mammals?* ①

5–3 (continued)

CONTENT DEVELOPMENT
Explain that Australia has many marsupials besides kangaroos and koalas, which are two of the most familiar. Australia's isolation as an island continent for millions of years allowed marsupials to survive there.

● ● ● ● **Integration** ● ● ● ●

Use the discussion of Australia's marsupials to integrate social studies into your science lesson.

INDEPENDENT PRACTICE
Section Review 5–3

1. Pouched mammals give birth to young that are not well developed. The young complete their development in a pouchlike structure in their mother's body. Kangaroos and koalas are two examples of pouched animals.
2. Marsupial.
3. An animal that eats only plant material.
4. Koalas and bears are both mammals,

Opossums

Have you ever heard the phrase "playing possum"? Do you know what it means? When in danger, opossums lie perfectly still, pretending to be dead. In some unknown way, this behavior helps to protect the opossum from its predators.

The opossum is the only pouched mammal found in North America. It lives in trees, often hanging onto branches with its long tail. Opossums eat fruits, insects, and other small animals.

Female opossums give birth to many opossums at one time. The females of one species of opossums, for example, give birth to as many as 56 offspring. Unfortunately, most of the newborn opossums do not survive the trip along their mother's abdomen to her pouch. Another species of female opossums produces newborns that are so tiny they have a mass of only 0.14 gram (about the mass of a small nail)!

5–3 Section Review

1. What are pouched mammals? Give two examples of these mammals.
2. What is another name for a pouched mammal?
3. What is a herbivore?

Critical Thinking—*Applying Facts*
4. Explain why it is incorrect to refer to the koala as a koala bear.

Endangered Mammals

Endangered species are of concern to almost everyone. Using reference materials in the library, develop an ongoing list of endangered mammals. Be sure to include an illustration of each mammal on your list. Keep your list up to date.

What suggestions can you make to help prevent other mammals from becoming endangered species?

5–4 Placental Mammals

Placental mammals give birth to young that remain inside the mother's body until their body systems are able to function independently. **Unlike the young of egg-laying and pouched mammals, the young of placental mammals develop more fully within the female.** The name for this group of mammals comes from a structure called a **placenta** (pluh-SEHN-tuh), which develops in females who are pregnant. Through the placenta, food, oxygen, and

Guide for Reading

Focus on these questions as you read.
▶ What are placental mammals?
▶ What are ten major groups of placental mammals?

C ■ 127

5-4 Placental Mammals

MULTICULTURAL OPPORTUNITY 5-4

Tell students the Cherokee legend of Awi Usdi, the Little Deer, which stresses the American Indians' respect for animals. In early times, people and animals could communicate with each other, and they lived in harmony. People killed animals only as needed for food and clothing. After the development of the bow and arrow, things changed, and some animals were slaughtered almost to extinction. The animals got together to see what they could do to remedy things. Awi Usdi, the Little Deer, had a suggestion. She would whisper in the ears of the hunters that they must first ask permission of the animals they were hunting and then ask for pardon from the spirit of the animal they had to kill for food. If the hunters failed to do so, Little Deer would cast a magic spell on them and cause them to be crippled. The next day, some hunters heeded the advice of Little Deer, and others did not. The ones who did not became crippled. According to the legend, that is why, to this day, American Indians hunt only animals that they need for their food and clothing and give thanks to the animals that they hunt.

ESL STRATEGY 5-4

Make sure that students remember what antonyms are (words that are opposite in meaning). Have them give the antonym for *herbivore* and list the names of five herbivores, five land-living carnivores, and five sea-living carnivores. Have them arrange the words in each list in alphabetical order.

---

but they are different kinds of mammals. A koala is a marsupial; it gives birth to young who must develop in a pouch. A bear is a placental mammal; it gives birth to young who have developed fully inside the mother's body.

REINFORCEMENT/RETEACHING

Monitor students' responses to the Section Review questions. If students appear to have difficulty with any of the questions, review the appropriate material in the section.

CLOSURE

▶ *Review and Reinforcement Guide*

At this point have students complete Section 5–3 in the *Review and Reinforcement Guide.*

TEACHING STRATEGY 5-4

FOCUS/MOTIVATION

Some students may have been present at the birth of puppies, kittens, or other

animals. Ask them to describe what they observed.

• **How was the offspring attached to its mother?** (By the umbilical cord.)

Explain that the other end of the umbilical cord is attached to the placenta. Show students several circular disks ranging in size from 0.5 to 20 centimeters in diameter. Point out that the placenta is circular in shape and varies in size according to the size of the mammal.

Figure 5–8 *Placental mammals, such as the deer mouse, give birth to young that remain inside the mother's body until they can function on their own. What is the placenta?* 1

wastes are exchanged between the young and their mother. Thus the placenta allows the young to develop for a much longer time inside the mother.

The time the young spend inside the mother is called the **gestation** (jehs-TAY-shuhn) **period.** The gestation period in placental mammals ranges from a few weeks in mice to as long as 18 to 23 months in elephants. In humans, the gestation period is approximately 9 months. After female placental mammals give birth to their young, they supply the young with milk from their mammary glands. You should remember that this is a characteristic of all mammals.

There are many groups of placental mammals, organized here according to how they eat, how they move about, or where they live. Of these groups, 10 are discussed in the sections that follow. 1

Insect-Eating Mammals

What has a nose with 22 tentacles and spends half its time in water? The answer is a star-nosed mole. As you can see from Figure 5–9, a star-nosed mole gets its name from the ring of 22 tentacles on the end of its nose. No other mammal has this

Figure 5–9 *Insect-eating mammals include the star-nosed mole, the hedgehog, and the pygmy shrew. The star-nosed mole (top right) lives in moist or muddy soil in eastern parts of North America. The hedgehog (bottom right), which is covered by a thick coat of spines, curls into a ball when threatened. The pygmy shrew (bottom left) must eat twice its mass in insects every day to survive.*

5–4 (continued)

CONTENT DEVELOPMENT

Point out that unlike egg-laying and pouched mammals' young, the young of placental mammals develop totally within the female. Explain that the placenta is the life-giving organ that is developed within the female mammal. Tell students that a placenta is a thick, membranous organ that connects the mother to the baby. Through the membranes, food and oxygen are carried to the baby from the mother, and metabolic wastes are removed.

● ● ● ● **Integration** ● ● ● ●

Use the discussion on how mammals are organized into groups to integrate classification concepts into your lesson.

FOCUS/MOTIVATION

Have students look at the photograph of the star-nosed mole.
● **How did this animal get its name?** (From the ring of tentacles at the end of its nose.)
● **How does this structure enable the star-nosed mole to catch food, even with poor eyesight?** (The tentacles help the mole find insects.)
● **Where might this mole be found?** (Underground.)

● **Look at the mole's feet. How are they adapted to the mole's lifestyle?** (The feet are specialized for burrowing in the ground or paddling in water.)

CONTENT DEVELOPMENT

Point out that the insect-eating group of placental mammals includes moles, shrews, and hedgehogs. Explain that moles eat grubs and worms. Shrews eat grubs, worms, mice, and other shrews. Hedgehogs eat almost all available inver-

structure. Each tentacle has very sensitive feelers, which enable a mole to find insects to eat and to feel its way around while burrowing under ground. Although star-nosed moles have eyes, the eyes are too tiny to see anything. They can only distinguish between light and darkness. A star-nosed mole spends one part of its day burrowing beneath the ground and the other part in the water.

In addition to moles, hedgehogs and shrews are also insect-eating mammals. Because it is covered with spines, a hedgehog looks like a walking cactus. When threatened by a predator, a hedgehog rolls up into a ball with only its spines showing. This action makes a hedgehog's enemy a little less enthusiastic about disturbing the tiny animal.

The pygmy shrew is the smallest mammal on Earth. As an adult, it has a mass of only 1.5 to 2 grams (about the mass of 10 small nails). Because shrews are so active, they must eat large amounts of food to maintain an adequate supply of energy. Shrews can eat twice their mass in insects every day!

Flying Mammals

Bats, which resemble mice, are the only flying mammals. In fact, the German word for bat is *Flieder-maus*, which means flying mouse. Bats are able to fly because they have skin stretched over their arms and fingers, forming wings. Other mammals, such as flying squirrels, do not really fly. They simply glide to the ground after leaping from high places.

Although some bats have poor eyesight, all bats have excellent hearing. While flying, a bat gives off high-pitched squeaks that people cannot hear. These squeaks bounce off nearby objects and return to the bat as echoes. By listening to these echoes, a bat knows just where objects are so that it can avoid them. Bats that hunt insects such as moths also use this method to locate their prey.

There are two main types of bats. Fruit-eating bats are found in tropical and desert areas, such as Africa, Australia, India, and the Orient. Insect-eating bats live almost everywhere.

Figure 5–10 *Bats are the only flying mammals. A mouse-eared bat swoops down on its prey—a moth—while the long-tongued bat shows the feature for which it was named.*

ACTIVITY WRITING

Useful Mammals

Visit your library to find out which mammals have benefited people. On a sheet of paper, make a list of these mammals. Next to each entry indicate the mammal's benefit to people. Discuss with your class whether the benefit listed for each mammal has been harmful, helpful, or neutral to the mammal. Was it necessary for people to use these mammals for their own survival?

C ■ 129

5-4 (continued)

GUIDED PRACTICE

Skills Development

Skills: Making observations, making comparisons

At this point have students complete the in-text Chapter 5 Laboratory Investigation: Examining Hair. Students will observe and describe the characteristics of hair.

FOCUS/MOTIVATION

Have students rub their tongues across the edge of their upper teeth.
• **What do you feel?** (Accept all logical answers.)

Point out that the center four teeth are called incisors and are used for biting. The longer, pointed teeth on either side of the incisors are called canines.

CONTENT DEVELOPMENT

Point out that all flesh-eating mammals, called carnivores, have sharp, pointed teeth called canines. Explain that canine teeth are used for tearing and shredding meat.

Tell students that carnivores are predators.

ACTIVITY
CALCULATING

The Fastest Runner

❶ The cheetah, the fastest land animal, can run at speeds of up to 100 kilometers per hour. How far can it run in a second? In a minute?

Flesh-Eating Mammals

The frozen ice sheets and freezing waters of the Arctic are home to walruses. They are able to live in this harsh environment because they have a layer of fat, called blubber, under their skin. Blubber keeps body heat in. In addition to their large size, walruses have another noticeable feature: long tusks. These tusks are really special teeth called canines. Walrus canines can grow to 100 centimeters in length!

Walruses do not use their canines (tusks) for tearing and shredding food. Instead, they use their canines to open clams and to defend themselves from their predators: polar bears. Walruses also use their tusks as hooks to help them climb onto ice.

Figure 5–11 *Flesh-eating mammals such as the walrus, the tiger, the coyote, and the grizzly bear are placed in the same group. All these mammals have large pointed teeth that help them tear and shred flesh. What are some other examples of flesh-eating mammals?* ❶

• **What is a predator?** (Accept all answers.)

Explain that a predator is an animal that kills and eats other animals. Point out that sea-mammal carnivores include the seal, walrus, and sea lion. Land carnivores are divided into three subcategories: (1) those that walk flat-footed on the soles of their feet, such as bears and raccoons; (2) those that walk on their toes, such as cats, dogs, wolves, and tigers; and (3) those that walk with a combination, partly on the toes and partly on the soles of the feet, such as skunks, otters, minks, and weasels.

GUIDED PRACTICE

Skills Development

Skill: Identifying patterns

Have students research and make drawings of the footprints of different carnivorous mammals. Have them identify the mammal and the subgroup of walking style on the back of the drawing.

Like a walrus, you too have canines. However, your canines are not as large as a walrus's. You have a total of four canines: two in your upper set of teeth and two in your lower. To find your upper canines, look in a mirror and first locate your eight incisors, or front teeth—four on the bottom and four on top. Incisors are used to bite into food. On either side of your top and bottom incisors is a tooth that comes to a point. These teeth are your canines.

All flesh-eating mammals, including walruses, are known as carnivores (KAHR-nuh-vorz). Some carnivores live on land; others, such as walruses, live in the sea. Land carnivores such as cats, dogs, weasels, lions, wolves, and bears and their relatives have muscular legs that help them to chase other animals. These mammals also have sharp claws on their toes to help them hold their prey.

Like walruses, seals are flesh-eating mammals that live in the sea. The ancestors of these sea carnivores once lived on land but have since returned to the ocean, where they feed on fishes, mollusks (soft-bodied invertebrates that have inner or outer shells), and sea birds. The flipperlike arms and legs of seals, which are so useful in getting around in the water, make moving from place to place on land quite difficult. Regardless of their struggle while on land, seals frequently leave the water.

Toothless Mammals

Although the name of this group indicates that its members do not have any teeth, there are some toothless mammals that actually do have small teeth. They include armadillos and sloths. The mammals in this group that actually have no teeth are the anteaters.

Unlike the spiny anteaters mentioned earlier in this chapter, the anteaters in this group do not lay eggs. As in all placental mammals, the young anteaters remain inside the female until they are more fully developed. However, both types of anteaters have something in common—a long sticky tongue that is used to catch insects. Look at Figures 5–5 and 5–12. Can you see any other similarities or differences between these two anteaters? ❷

C ■ 131

Answers

❶ Answers will vary. Examples include lions, dogs, cats, and wolves. (Applying concepts)

❷ Spiny anteaters and anteaters both have strong claws for digging. Spiny anteaters have sharp spines; anteaters have heavy fur coats. (Interpreting illustrations)

Integration

❶ Mathematics

FACTS AND FIGURES
TOOTHLESS ANCESTORS

Some prehistoric "toothless" mammals, or edentates, were enormous. Certain giant ground sloths grew to the size of modern elephants; they were about 6 meters long. An armored, armadillolike edentate was about 5 meters long and carried a 3-meter-shell on its back.

INTEGRATION
MEDICINE

Armadillos are thought to be the only nonhuman animals that can catch Hansen's disease, or leprosy. Nine-banded armadillos are often used in research of this disease because the four offspring in a litter are genetically identical.

Explain that only the true anteater is actually toothless. Although the sloth and the armadillo have no teeth in the front, they do have poorly developed back teeth.

Tell students that true anteaters have large, sickle-shaped claws on their front paws, which they use for defense and for opening the nests of ants and termites. The largest anteater, the giant anteater, is about 2 meters long and has a mass of about 35 kilograms. The young giant anteater is dependent on its mother until it is about two years old. For the first year of life, the young anteater often rides on its mother's back.

Have other students observe and identify the type of walking style using the drawings.

CONTENT DEVELOPMENT

Explain that many mammals classified as carnivores also include vegetable matter in their diets. In fact, many bears are largely herbivores that eat meat only on occasion. They are still classified as carnivores, however, because of the presence of the enlarged canine teeth that make it possible for them to eat flesh.

REINFORCEMENT/RETEACHING

▶ *Activity Book*

Students who need practice with classifying animals should complete the chapter activity Find the Oddball. Ask students to explain what three of the animals pictured have in common.

CONTENT DEVELOPMENT

Point out that great or true anteaters, sloths, and armadillos belong to a group of placental mammals called toothless.

ELEPHANTS

Elephants were at one time found all over the world except in Australia, New Zealand, and Antarctica. The largest one known was the imperial mammoth that flourished in North America, Europe, Asia, and Africa. During the ice age, these elephants were a staple part of the diets of humans, and it is possible they were hunted into extinction. Today, only about 1 million African elephants are estimated to live in the wild, and only about 50,000 Asiatic elephants remain.

The heaviest land animal alive today is the African elephant. Males can have a mass of up to 7500 kilograms, which is equal to that of about eight compact cars. Such elephants can be 7 meters long, trunk to tail, and 4 meters tall at the shoulder. These animals eat only vegetation, which makes them herbivores.

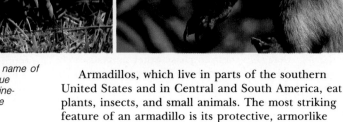

Figure 5–12 *Just as the name of their group implies, the true anteater (right) and the nine-banded armadillo (left) are toothless mammals.*

Figure 5–13 *The larger ears of the African elephant (bottom) distinguish it from the Indian elephant (top).*

Armadillos, which live in parts of the southern United States and in Central and South America, eat plants, insects, and small animals. The most striking feature of an armadillo is its protective, armorlike coat. In fact, the word armadillo comes from a Spanish word that means armored. The nine-banded armadillo is the only toothless mammal found as far north as the United States.

Another type of toothless mammal is the sloth. There are two kinds of sloths: the two-toed sloth and the three-toed sloth. Sloths, which feed on leaves and fruits, are extremely slow-moving creatures. They spend most of their lives hanging upside down in trees. You may be amazed to learn that sloths can spend up to 19 hours a day resting in this position.

Trunk-Nosed Mammals

As the elephant enters the river, it holds its trunk high in the air. Little by little, the water creeps up the elephant's body. Will it drown? The answer comes a few seconds later as the huge animal actually begins to swim!

To an observer on the shore, nothing can be seen of the elephant except its trunk, through which air enters on its way to the lungs. The trunk is the distinguishing feature of all elephants. Elephant trunks are powerful enough to tear large branches from trees. Yet they are agile enough to perform delicate movements, such as picking up a peanut thrown to them by a child at a zoo. These movements are

---

5–4 (continued)

CONTENT DEVELOPMENT

Point out that the "hair" of the armadillo is modified into heavy, overlapping, bony scales. When threatened, armadillos roll up so that only the scales are exposed. Then they wait for the enemy to go away. Tell students that armadillos always give birth to identical twins or quadruplets.

Tell students that sloths are often distinctly greenish in color. This color is due to symbiotic blue-green algae that grow in the animal's hair. The algae help to camouflage the sloth. Sloths also have a symbiotic relationship with cellulose-digesting bacteria, which live in the sloth's stomach.

● ● ● ● **Integration** ● ● ● ●

Use the discussion of the armadillo's name to integrate language arts concepts into your science lesson.

GUIDED PRACTICE

▶ *Laboratory Manual*

Skills: Making observations, manipulative, making comparisons, relating concepts

Students may now complete the Chapter 5 Laboratory Investigation in the *Lab-*

oratory Manual called The Most Intelligent Mammals. They will study human behavior, including learned and unlearned responses and the role of reflexes, conditioned responses, habits, trial-and-error learning, and reasoning.

CONTENT DEVELOPMENT

Point out that an elephant's trunk is a modification of the upper lip and nose. While the trunk is powerful enough to tear branches from trees, it is also a sen-

sitive organ of smell and touch. Small hairs at the tip enable the elephant to pick up very small objects. Point out that elephants use their trunks for many purposes. In addition to serving as a snorkel for swimming, it is used to suck up water and squirt it into the mouth. The trunk is also used to spray water over the body to cool off. Elephants touch trunks to greet one another, and they raise them as a sign of aggression.

made possible by the action of 40,000 muscles, which are controlled by a highly developed brain.

Elephants are the largest land animals. There are two kinds of elephants: African elephants and Asian elephants. As their names suggest, African elephants live in Africa and Asian elephants in Asia, especially in India and Southeast Asia. Although there are a number of differences between the two kinds of elephants, the most obvious one is ear size. The ears of African elephants are much larger than those of their Asian cousins.

Hoofed Mammals

What do pigs, camels, horses, and rhinoceroses have in common? Not much at first glance. In fact, they could not look more different. Yet if you were to take a closer look at their feet, you would see that they all have thick hoofs.

If you could take an even closer look at several hoofed mammals, you would discover that some have an even number of toes, whereas others have an odd number of toes. Pigs, camels, goats, cows, deer, and giraffes—the tallest of all mammals—have an even number of toes. Horses, rhinoceroses, zebras, hippopotamuses, and tapirs have an odd number of toes.

Hoofed mammals such as pigs, cows, deer, and horses are important to humans and have been so for thousands of years. These hoofed mammals provide humans with food and clothing, as well as with a means of transportation.

Migration of Mammals

Certain mammals migrate, or move, to places that offer better living conditions. Using posterboard and colored pencils, draw maps that trace the migration patterns of the following mammals: North American bat, African zebra, American elk, and gray whale. Display these maps on a bulletin board at school. ❷

■ Why do you think each type of mammal migrates?

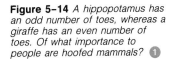

Figure 5–14 *A hippopotamus has an odd number of toes, whereas a giraffe has an even number of toes. Of what importance to people are hoofed mammals?* ❶

C ■ 133

ACTIVITY
DISCOVERING
MIGRATION OF MAMMALS

Discovery Learning

Skills: Making comparisons, relating concepts, recording information, diagramming

Materials: posterboard, colored pencils

Migration studies at this grade level are usually limited to studies of birds; however, the migration patterns of certain mammals are also interesting. The maps drawn by students can be referred to as you teach about the flying, hoofed, and water-dwelling mammals involved. The reasons why animals migrate will vary. Some return to certain places to breed; some move to places where food is more plentiful.

Integration: Use this Activity to integrate fine arts skills into your science lesson.

FOCUS/MOTIVATION

Show students several pictures of hoofed mammals such as a horse, cow, deer with antlers, giraffe, or bison.

• **What do you predict these placental mammals have in common?** (Accept all logical answers.)

• **How are some of these mammals helpful to people?** (Accept all logical answers.)

CONTENT DEVELOPMENT

Point out that the even-toed mammals, such as the cow, sheep, pig, and hippopotamus, are subdivided into two groups according to their eating habits. There are cud-chewing and noncud-chewing even-toed mammals.

• **Which mammals do you know of that are cud-chewing?** (Most students will say cows.)

Explain that cows, sheep, goats, giraffes, and deer are cud-chewing mammals. They have four divisions to their stomachs. They eat a large amount of food that quickly passes to the first stomach division. Later, it is brought back to the mouth, where it is chewed and passed to the second division of the stomach, where digestion begins. Point out that in noncud-chewing mammals, the food goes directly to the stomach to start digestion.

UNGULATES

The odd-toed and even-toed hoofed mammals, or ungulates, are thought to have diverged from a common ancestor about 60 million years ago. The two orders of hoofed mammals, however, share many similarities in form and function.

Most hoofed animals are adapted for swift running on open ground. Their muscular legs are of equal length and are designed for back-and-forth movement. The limited movement in the leg joints allows for fast running but does not permit activities such as climbing and digging. The number of toes is reduced, and the foot is elevated so that all of the body weight is on the tips of the toes. The toes are covered by hard, thick hooves, which are actually enlarged nails.

In addition to being suited to running, most hoofed mammals are suited to low-protein, high-fiber diets of plant matter. A hoofed mammal's teeth have large grinding surfaces, which break down cellulose plant-cell walls to release the nutritious cell contents. The digestive system is also modified for handling large quantities of cellulose.

Figure 5–15 *The most numerous mammals are the gnawing mammals. Two of the four special incisors that help gnawing mammals chew on hard objects such as wood and nuts can be seen in the photographs of the nutria (center) and the porcupine (bottom). Although the incisors of the chipmunk (top) are concealed by a mouthful of seeds, it too is a gnawing mammal.*

134 ■ C

Gnawing Mammals

Hardly a day goes by that people in both the country and the city do not see a gnawing mammal. There are more gnawing mammals than there are any other type of mammal on the Earth. Gnawing mammals are more commonly known as rodents.

Among the rodents are such animals as squirrels, beavers, chipmunks, rats, mice, and porcupines. As you might expect, the characteristic that places these animals in the same group is the way they eat: They gnaw, or nibble. Gnawing mammals have four special incisors. These teeth are chisellike and continue to grow throughout the animal's lifetime. Because rodents gnaw on hard objects such as wood, nuts, and grain, their teeth get worn down as they grow. If this were not the case, a rodent's incisors would become so long that the animal would not be able to open its mouth wide enough to eat.

Some rodents, especially rats and mice, compete with humans for food. They eat the seeds of plants and many other foods used by people. Rodents are responsible for spreading more than 20 diseases, including bubonic plague—which is actually transmitted to people by the bite of a flea that lives on rats. Although it is no longer considered a serious disease, bubonic plague caused the deaths of 25 million people in Europe between the years 1300 and 1600.

Rodentlike Mammals

Rabbits, hares, and pikas (PIGH-kuhz) belong to the group known as rodentlike mammals. Like rodents, these mammals have gnawing teeth. Unlike rodents, however, they have a small pair of grinding teeth behind their gnawing teeth. Rodentlike mammals move their jaws from side to side as they chew their food, whereas rodents move their jaws from front to back.

Rabbits and hares have long hind legs that are used for quick movement and flight from danger. Some larger hares can reach speeds of up to 80 kilometers per hour! In addition, rabbits and hares have large eyes that enable them to remain active during the night.

5-4 (continued)

CONTENT DEVELOPMENT

Have students observe the rodents in Figure 5–15.
• **What traits do these rodents have in common?** (Lead students to answer that rodents are gnawing animals.)
• **The gnawing mammals are the most numerous of all. What are some different ones that you know of?** (Many answers are possible. As correct animals are named, list them on the chalkboard.)
• **What teeth do these mammals use for gnawing?** (The incisors are the gnawing teeth.)
• **How are a rodent's incisors different from yours?** (They are chisel-shaped and continue to grow throughout life.)

● ● ● ● **Integration** ● ● ● ●

Use the discussion of bubonic plague in Europe in the Middle Ages to integrate concepts of disease into your lesson.

INDEPENDENT PRACTICE

▶ *Activity Book*

Students can practice observing similarities and differences among animals in the chapter activity called Classifying Animals. Students are asked to identify one

shared characteristic and one difference in a group of four animals.

CONTENT DEVELOPMENT

Explain that rabbits, hares, and pikas were once classified as rodents because they also have chisellike incisors. Because of other differences, however, they are now placed in another group.
• **How are the teeth of rabbits and hares different from rodent teeth?** (Smaller

How are hares and rabbits different? In terms of appearance, hares tend to have longer legs and longer ears than rabbits do. Hares also live on their own on the surface of the ground. Rabbits are born in burrows and are helpless for the first few weeks of their lives.

Pikas, which have large rounded ears and short legs, are not well known because they live high up in the mountains or below ground in burrows. If you are interested in seeing a pika, look at Figure 5–16.

Water-Dwelling Mammals

"Thar she blows!" is the traditional cry of a sailor who spots the fountain of water that a whale sends skyward just before it dives. Although sailors of the past recognized this sign of a whale, they had no idea that this sea animal was a mammal, not a fish.

Whales, porpoises, dolphins, dugongs, and manatees are water-dwelling mammals. Although they live in water most or all of the time, they have lungs and breathe air. They feed their young with milk and have hair at some time in their life.

Whales, dolphins, and porpoises spend their entire lives in the ocean and cannot survive on land. Dugongs and manatees live in shallower water, often in rivers and canals. Because of their large size, it is difficult for dugongs and manatees to move around on land. However, they do so for short periods of time when they become stranded.

Figure 5–16 *The pika (top) and the rabbit (bottom) belong to the group known as the rodentlike mammals.*

Figure 5–17 *Two examples of mammals that live in water are the manatee (left) and the humpback whale (right). What are some other examples of water-dwelling mammals?* ❶

FACTS AND FIGURES
MAMMAL SIZE

The largest mammal of all is the blue whale, whose head-to-tail length is about 27 meters. The largest whale ever caught was 33.58 meters long. The whale can have a mass of up to 140,000 kilograms. The largest land animal, the African elephant, is tiny by comparison. Its trunk-to-tail length is about 7 meters, and it may have a mass of about 7500 kilograms.

include the breeds of mammals that cannot survive out of water. Water-dwelling mammals breathe air and normally come to the surface of the water to breathe.

Explain that some of the water-dwelling mammals, such as the dugongs and manatees, live in shallow water. Point out that occasionally these shallow-water mammals become stranded on land and must move back to water for survival.

REINFORCEMENT/RETEACHING

▶ *Activity Book*

Students who need more practice identifying the characteristics of animals should complete the chapter activity called Scoring Vertebrates. Students will observe and compare the characteristics of five vertebrate animals to discover which one has the most advanced characteristics.

grinding teeth are behind the gnawing teeth.)

• **In what other way are these animals different from rodents?** (The rodentlike mammals move their jaws from side to side as they chew.)

FOCUS/MOTIVATION

Media and Technology

Have students examine the relationships between whales, humans, and the global ecosystem through the use of the Interactive Videodisc entitled In the Company of Whales. Students should begin their exploration of whales by taking a closer look at the biological features of the largest animals that have ever lived on Earth.

CONTENT DEVELOPMENT

Although many mammals live in water, such as the beaver, muskrat, and hippopotamus, the water-dwelling mammals

CLOTHING AND THE ENVIRONMENT

Throughout history, mammal skins have been used for clothing humans. When the ice age ended about 10,000 years ago, people on the North American continent used skins from saber-toothed cats, bears, wolves, and woolly mammoths to keep them warm. Native Americans later used the pelts of elk, deer, buffalo, and bear for clothing. Early colonists continued these uses, and the European settlers of North America brought with them a tradition of making clothing out of spun fibers such as sheep's wool.

Modern technology has created synthetic clothing materials such as nylon and polyester. Many of these synthetic materials are derived from nonrenewable natural resources such as petroleum. Today, Americans wear clothing made from these synthetic fibers as well as fur coats, leather goods, and woolen products.

Have students analyze the different kinds of clothing they wear according to the natural resources from which they were derived. What kinds of impacts do their clothing preferences have on the environment? In their opinion, which kinds of clothing have the least damaging impact on the environment? Should any products or practices related to the clothing industry be prohibited?

Figure 5–18 *Of all the animals, primates, such as the gibbon (top), the chimpanzee (center), and the orangutan (bottom), have the most highly developed brain and the most complicated behaviors. What are some other characteristics of primates?* ❶

Primates

On a visit to your local zoo, you come upon a crowd of people gathered in front of one of the cages. Hurrying over to see what the excitement is about, you hear strange noises. Carefully you make your way to the front of the crowd, where you see what is causing all the commotion. A family of chimpanzees is entertaining the crowd by running and tumbling about in their cage. The baby chimpanzee comes to the front of the cage and extends its hand to you. You are amazed to see how much like your hand the chimpanzee's hand looks. But it is really no wonder they are similar. After all, the chimpanzee— along with the gibbon (GIHB-uhn), orangutan (oh-RANG-oo-tan), and gorilla—is the closest mammal in structure to humans.

These mammals, along with baboons, monkeys, and humans, belong to the same group—the primates. All primates have eyes that face forward, enabling the animals to see depth. Primates also have five fingers on each hand and five toes on each foot. The fingers are capable of very complicated movements, such as the ability to grasp objects.

Primates also have large brains and are considered the most intelligent mammals. Chimpanzees can be taught to communicate with people by using sign language. Some scientists have reported that chimpanzees can use tools, such as when they use twigs to remove insects from a log. Humans, on the other hand, are the only primates that can make complex tools.

5–4 Section Review

1. What are placental mammals? What is the function of a placenta?
2. What is a gestation period?
3. How do a carnivore and a herbivore differ?
4. List ten groups of placental mammals and give an example of each group.

Connection—*Ecology*
5. Elephants often use their tusks to strip the bark from trees. How might this action harm the environment?

5–4 (continued)

CONTENT DEVELOPMENT

Mention that primates are adapted to a wide range of habitats and thus show a wide range of posture and methods of locomotion. Some of the primitive primates, like lemurs and tarsiers, resemble squirrels, and like squirrels, they scamper about on tree branches on all fours. Monkeys are somewhat more up-right, but they also run along branches on all fours. Some apes, like gibbons and orangutans, spend most of their time in trees, swinging from limb to limb with their arms. Chimpanzees and gorillas spend considerable time on the ground, and they can walk short distances on their hind legs. But most of the time they also use the knuckles of their hands for walking. Humans are the only primates that walk completely upright.

ENRICHMENT

▶ *Activity Book*

Students will be challenged by the chapter activity Classifying Mammals in which they compare the features of mammals and correct a chart of observations.

INDEPENDENT PRACTICE

Section Review 5–4

1. Placental mammals give birth to young that remain inside the mother's body until their body systems are able to

Do You Hear What I Hear? ❶

The question posed in the title is really not so easily answered. If you were a cat, dog, porpoise, or bat—and you could speak—you would have four different answers to this question. Why? Each one of these mammals is capable of hearing sounds other animals cannot!

Sound is a form of energy that is produced when particles vibrate. The number of particle vibrations per second, or the frequency, is an important characteristic of sound. The ear can respond only to certain frequencies. For example, the normal human ear is capable of detecting sounds that have between 20 and 20,000 vibrations per second. Sounds with frequencies higher than 20,000 vibrations per second are called *ultrasonic sounds* (the prefix *ultra*-means beyond). Ultrasonic sounds cannot be heard by the human ear.

Some animals can hear ultrasonic sounds, however. If you have ever used a dog whistle, you know that when you blow on it, your dog comes running even though you cannot hear any sound. Dogs can hear sounds that have frequencies up to 25,000 vibrations per second. Cats can hear sounds with frequencies up to 65,000 vibrations per second. The frequency limit for porpoises is 150,000 vibrations per second. And bats can hear ultrasonic sounds with frequencies up to 200,000 vibrations per second. Bats not only hear ultrasonic sounds, they also produce ultrasonic sounds. They use the echoes to avoid bumping into things and to locate prey such as moths. This process of navigation is called *echolocation*. The echolocation of many bats is so efficient that they can make last-minute swerves in order to intercept a moth that has changed its course!

function independently. The placenta, which develops in pregnant females, allows food, oxygen, and wastes to be exchanged between the young and the mother.

2. The time the young spend inside the mother.

3. Carnivores are flesh-eaters; herbivores eat plant material.

4. Insect-eating—hedgehog; flying—bat; flesh-eating—bear; toothless—armadillo; trunk-nosed—elephant; hoofed—horse; gnawing—squirrel; rodentlike—rabbit; water-dwelling—whale; primates—gorilla. Examples will vary.

5. Elephants can destroy the trees because trees without bark are more susceptible to disease and insects and therefore are more likely to die.

REINFORCEMENT/RETEACHING

Review students' responses to the Section Review questions. Reteach any material that is still unclear, based on students' responses.

CLOSURE

▶ *Review and Reinforcement Guide*
Students may now complete Section 5–4 in the *Review and Reinforcement Guide*.

Laboratory Investigation

EXAMINING HAIR

BEFORE THE LAB

1. Assemble the materials needed for each student.

2. Purchase some inexpensive combs. You can ask students to furnish their own, but some will forget. For hygienic reasons, students should not be allowed to share combs or brushes.

3. Check the dilutions on the methylene blue. Too strong a solution makes it more difficult to see the cells.

4. Prepare methylene blue stain by diluting 10 milliliters of stock solution with 90 milliliters of stain. Stock solutions of methylene blue can be prepared by adding 1.48 grams of methylene blue powder to 100 milliliters of 95 percent ethyl alcohol and letting it stand for two days. During this time, stir frequently. After two days, filter.

PRE-LAB DISCUSSION

Review the body coverings of all vertebrates, including the mammals.

Review the process involved in correctly preparing a wet-mount slide, reminding students that they are using methylene blue instead of water.

• **What is the purpose of this laboratory investigation?** (To discover the characteristics of hair.)

• **What is significant about hair?** (It is the body covering for all mammals.)

• **What is the function of hair?** (It is part of a mammal's system for conserving body heat.)

Laboratory Investigation

Examining Hair

Problem

What are the characteristics of hair?

Materials *(per student)*

medicine dropper	coverslip
methylene blue	microscope
glass slide	electric light
comb or brush	hand lens
scissors	

Procedure

1. Using the medicine dropper, put 2 drops of methylene blue in the center of a clean glass slide. **CAUTION:** *Be careful when using methylene blue because it may stain your skin and clothing.*

2. Comb or brush your hair vigorously to remove a few loose hairs.

3. From your comb or brush, select two hairs that each have a root attached. The root is the small bulb-shaped swelling at one end of the hair.

4. Using the scissors, trim the other end of each hair so it will be short enough to fit on the glass slide. Place the trimmed hairs in the drops of methylene blue on the slide. Cover the slide.

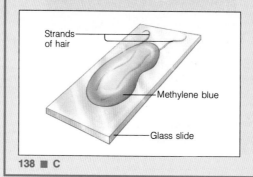

Strands of hair

Methylene blue

Glass slide

138 ■ C

5. Use the low-power objective of the microscope to locate the hairs. Then switch to the high-power objective and focus with the fine adjustment. Make a sketch of the hair strand.

6. Turn the fine adjustment toward you (counterclockwise) to focus on the upper surface of the hair. At one point in your focusing, the hair will appear to be covered by overlapping structures that look like shingles. Draw these structures, which are actually cells.

7. Under bright light, use the hand lens to examine a portion of your skin that does not seem to be covered with hair. (Do not examine the palms of your hands or the soles of your feet.)

Observations

1. Describe the appearance of a strand of hair under the microscope.

2. Describe the appearance of cells covering the strand of hair.

3. What did you observe when you examined the surface of your skin under bright light?

Analysis and Conclusions

1. Based on your observations, what is hair?

2. **On Your Own**   Obtain hair from several of the pets mentioned in the Journal Activity at the beginning of the chapter. Prepare these hair samples as you did your hair sample. Compare the various hair samples. Are they about the same thickness as your hair when observed with the unaided eye? When observed with the microscope? What other similarities or differences do you observe?

TEACHING STRATEGIES

1. Students should be monitored to ensure that they are carefully following the directions.

2. Some students may need help with placing the hairs on the slides. You may need to help them find and focus on the cells themselves.

DISCOVERY STRATEGIES

Discuss how the investigation relates to the chapter ideas by asking open questions similar to the following:

• **What is important about hair and animals?** (Hair is one of the distinguishing features of mammals—relating.)

• **Why is hair important to mammals?** (It is part of their system for conserving body heat—relating, concluding.)

• **Why can something be learned about mammalian hair by looking at human hair?** (Humans are mammals, so human

Study Guide

Summarizing Key Concepts

5–1 What Is a Mammal?

▲ Mammals are warmblooded vertebrates that have hair or fur and that feed their young with milk produced in mammary glands.

▲ Mammals use their lungs to breathe. Their circulatory system consists of a four-chambered heart and blood vessels. Paired kidneys filter nitrogen-containing wastes from the blood in the form of urea. Urea, water, and other wastes form urine, which is stored in the urinary bladder.

▲ Mammals have well-developed brains and senses, and they have internal fertilization.

▲ Mammals can be placed into three basic groups depending on the way in which they reproduce. These groups are the egg-laying mammals, the pouched mammals, and the placental mammals.

5–2 Egg-Laying Mammals

▲ Mammals that lay eggs are called egg-laying mammals, or monotremes.

▲ The duckbill platypus and the spiny ant-eater are examples of egg-laying mammals.

5–3 Pouched Mammals

▲ Unlike monotremes, pouched mammals do not lay eggs. Instead, they give birth to young that are not well developed. Thus the young must spend time in a pouchlike structure in their mother's body. Mammals that have pouches are called marsupials.

5–4 Placental Mammals

▲ Placental mammals give birth to young that remain inside the mother's body until their body systems are able to function independently. The placenta is a structure through which food, oxygen, and wastes are exchanged between the young and their mother.

▲ Insect-eating mammals include moles, hedgehogs, and shrews.

▲ Bats are the only flying mammals.

▲ Flesh-eating mammals, or carnivores, include sea-living animals such as walruses and seals. Land-living carnivores include any members of the dog, cat, and bear families.

▲ Toothless mammals include those animals that lack teeth or have small teeth.

▲ The only members of the trunk-nosed mammal group are the elephants.

▲ Hoofed mammals are divided into the group that has an even number of toes on each hoof and the group that has an odd number of toes.

▲ Gnawing mammals, such as beavers, chipmunks, rats, mice, and porcupines, have chisellike incisors for chewing.

▲ Rabbits, hares, and pikas are examples of rodentlike mammals.

▲ Mammals such as whales, dolphins, porpoises, dugongs, and manatees are water-dwelling mammals.

▲ Humans, monkeys, and apes are primates.

Reviewing Key Concepts

Define each term in a complete sentence.

5–1 What Is a Mammal?
egg-laying mammal
pouched mammal
placental mammal

5–4 Placental Mammals
placenta
gestation period

Part 1

Have students exchange slides and note the similarities and differences in the appearance of the hair samples, both with the unaided eye and the microscope.

Part 2

Have students make slides of bits of fingernails and compare their observations of these slides to their observations of their hair slides.

hair is a type of mammalian hair—inferring, relating.)

OBSERVATIONS

1. The inside appears to be made up of parallel layers of some uniform material. The outside is covered with overlapping layers of thin cells.

2. The cells are irregularly shaped and overlapping.

3. Even areas that appear to have no hair are covered with fine hairs.

ANALYSIS AND CONCLUSIONS

1. Answers will vary. Hair is a tube of keratin surrounded by dead epidermal cells.

2. Answers to the questions will vary, depending on the type of hair observed.

Chapter Review

ALTERNATIVE ASSESSMENT

The *Prentice Hall Science* program includes a variety of testing components and methodologies. Aside from the Chapter Review questions, you may opt to use the Chapter Test or the Computer Test Bank Test in your *Test Book* for assessment of important facts and concepts. In addition, Performance-Based Tests are included in your *Test Book*. These Performance-Based Tests are designed to test science process skills, rather than factual content recall. Since they are not content dependent, Performance-Based Tests can be distributed after students complete a chapter or after they complete the entire textbook.

CONTENT REVIEW

Multiple Choice

1. b
2. a
3. d
4. c
5. d
6. b
7. c
8. c
9. b
10. c

True or False

1. F, vertebrates
2. T
3. F, monotreme
4. T
5. T
6. F, rodentlike
7. T
8. T

Concept Mapping

Row 1: Amphibians, Birds, Mammals
Row 2: Monotremes, Placental mammals

CONCEPT MASTERY

1. Mammals are warmblooded vertebrates that have hair or fur and that feed their young with milk produced in mammary glands. Monotremes—lay eggs. Marsupials—young develop in pouches. Placentals—young develop within the body of the mother and are fed through the placenta.

2. An anteater is a placental mammal; a spiny anteater is a monotreme.

3. Kangaroo, opossum, koala. Their young are born in an immature state and complete their development inside the mother's pouch.

4. Female mammals have mammary glands, which produce milk to feed the young for a time after they are born. This characteristic is used most often to classify an animal as a mammal.

5. Insect-eating, flying, flesh-eating, toothless, trunk-nosed, hoofed, gnawing, rodentlike, water-dwelling, primates. Ex-

amples should be similar to those given in the text.

6. Whales have lungs and breathe air. They have hair and feed their young with milk. These are all characteristics of mammals, not fishes.

7. Carnivores often have keen senses of sight and smell to help them find their prey. They may be swift runners or have sharp claws for attacking prey. Mammals that are carnivores often have sharp teeth, particularly canines, for biting and tearing their prey.

Content Review

Multiple Choice

Choose the letter of the answer that best completes each statement.

1. All mammals have
   a. pouches.   c. feathers.
   b. hair.   d. fins.
2. The kangaroo is a(n)
   a. pouched mammal.
   b. egg-laying mammal.
   c. placental mammal.
   d. gnawing mammal.
3. The only North American marsupial is the
   a. kangaroo.   c. platypus.
   b. koala.   d. opossum.
4. Young mammals that develop totally within the female belong to the group called
   a. egg-laying mammals.
   b. pouched mammals.
   c. placental mammals.
   d. marsupial mammals.
5. Which is an insect-eating mammal?
   a. whale   c. bear
   b. elephant   d. mole
6. Which teeth are used to tear and shred food?
   a. carnivores   c. incisors
   b. canines   d. herbivores
7. Which is an example of a toothless mammal?
   a. skunk   c. armadillo
   b. mole   d. camel
8. The largest land animal is the
   a. blue whale. c. elephant.
   b. rhinoceros. d. giraffe.
9. An example of a water-dwelling mammal is the
   a. spiny anteater. c. shrew.
   b. dolphin.   d. elephant.
10. To which group of mammals do humans belong?
    a. insect-eating mammals
    b. rodents
    c. primates
    d. carnivores

True or False

If the statement is true, write "true." If it is false, change the underlined word or words to make the statement true.

1. Mammals are <u>invertebrates</u>.
2. Mammals are <u>warmblooded</u> animals.
3. The duckbill platypus is an example of a <u>marsupial</u>.
4. Animals that eat only plants are called <u>herbivores</u>.
5. The <u>gestation period</u> is the time the young of placental mammals spend inside their mother.
6. Rabbits are <u>gnawing</u> mammals.
7. Carnivore means <u>flesh eater</u>.
8. Humans are <u>primates</u>.

Concept Mapping

Complete the following concept map for Section 5–1. Refer to pages C6–C7 to construct a concept map for the entire chapter.

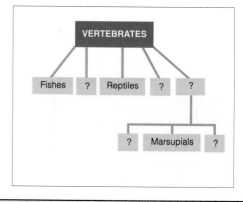

Concept Mastery

Discuss each of the following in a brief paragraph.

1. What are the characteristics of mammals? What are the three groups of mammals? How do they differ from one another?
2. What is the difference between an anteater and a spiny anteater?
3. Name two pouched mammals. Explain how their young develop.
4. What are mammary glands?
5. List ten groups of placental mammals and give an example of each.
6. Why are whales considered mammals and not fishes?
7. What features make carnivores good predators?
8. Why are primates considered the most intelligent mammals?

Critical Thinking and Problem Solving

Use the skills you have developed in the chapter to answer each of the following.

1. **Relating facts** Which group of mammals is most similar to birds? Explain your answer.
2. **Making charts** Prepare a chart with three columns. In the first column, list the ten groups of mammals that you learned about in this chapter. In the second column, list the characteristics of each group. And in the third column, list at least two examples of each group.
3. **Applying concepts** Why do whales usually come to the surface of the ocean several times an hour?
4. **Making inferences** Many species of hoofed mammals feed in large groups, or herds. What advantage could this behavior have for the survival of these mammals?
5. **Making generalizations** What is the relationship between how complex an animal is and the amount of care the animal gives to its young? Provide an example to support your answer.
6. **Designing an experiment** Design an experiment that will test the following hypothesis: Chimpanzees are able to understand the meaning of the spoken words for the numbers 1 through 10.
7. **Using the writing process** Imagine that you are a writer for a wildlife magazine. Your boss tells you that she has decided not to include mammals in the magazine because she finds them dull. Write a memo to change her mind. Include the characteristics that separate mammals from other animals.

6. Experiments should be logical and well thought out. Answers will vary.
7. Check all memos for clarity and correctness. Listing some of the wide range of mammals, from bats to whales, and their interesting lifestyles should convince even the most recalcitrant boss that mammals are indeed interesting. Offering the boss a mirror should convince her that she will find at least one mammal interesting.

KEEPING A PORTFOLIO

You might want to assign some of the Concept Mastery and Critical Thinking and Problem Solving questions as homework and have students include their responses to unassigned questions in their portfolio. Students should be encouraged to include both the question and the answer in their portfolio.

ISSUES IN SCIENCE

The following issue can be used as a springboard for discussion or given as a writing assignment:

At the beginning of the eighteenth century, there were about 9 million harp seals in the North Atlantic Ocean. Overhunting once drove the harp seal to the edge of extinction, but strict hunting laws allowed the population to recover. Today, harp seals have a population of about 4 million, which is growing by about 5 percent each year.

For the first two weeks of life, young harp seals have beautiful white coats. In the past, large numbers of seal pups have been killed for their fur, usually by clubbing them to death in the presence of their mothers. Conservation and humane groups strongly oppose the killing of the seals. They think that killing young seals for their fur is cruel and may have a negative effect on the seal population and the environment. Seal hunters, facing economic difficulty, want permission to kill greater numbers of young seals. They believe that the seal population is no longer in danger and that the young seals can be killed humanely. Should seal hunters be permitted to harvest greater numbers of seal pups? What is your opinion and why?

8. Primates have large brains. Also, they can use tools and have been taught to communicate with people using a sign language.

CRITICAL THINKING AND PROBLEM SOLVING

1. Answers will vary. Students will probably think that monotremes are the most similar to birds because monotremes lay eggs. Others may think that bats are the most similar because they fly.
2. Check charts to make sure they reflect the information presented in the chapter.
3. Whales have to come to the surface to breathe.
4. Hoofed animals may feed in herds because they can ward off predators when in groups or because an individual in a group is less likely to be preyed upon than one that is alone.
5. In general, animals that are more complex provide more care for their young than less complex animals provide. Students' examples will vary.

ADVENTURES IN SCIENCE

MICHAEL WERIKHE
SAVING THE RHINO—
ONE STEP AT A TIME

Background Information

The rhinoceroses are ungulates, or hoofed animals, and behave in many ways typical of ungulates. Like all other hoofed animals, they are plant-eaters, or herbivores. Alert to danger through keen senses, the rhinoceroses will usually run rather than stand and fight, and they wallow in mud just as pigs do because they cannot sweat.

The black rhinoceros of Africa feeds on leaves and twigs from a number of bushes. It has also been known to eat fruit that has fallen from trees. During the rainy season, the rhino browses for food in a wide area, but during the dry season, it always stays close to water. The white rhinoceros grazes on the African grasslands. Feeding on grass, it often eats in the cool of the morning and evening and rests during the hotter daytime hours.

Both the black and the white rhinoceroses have been threatened with possible extinction, but conservation efforts are changing that fortune. Through the conservation efforts of individuals, groups, and governments, the populations of the African rhinoceroses are not decreasing as rapidly now. Michael Werikhe's campaign is responsible in part for helping to save the rhino. Werikhe recognizes that conservation of the rhino and other threatened species is the responsibility of individuals as well as groups. As he noted in a book foreword, "My rhino campaign is only a symbol of conservation. . . . [t]he rhino will live or die because of us."

GAZETTE

MICHAEL WERIKHE
Saving the Rhino—One Step at a Time

There are few animals as impressive as a rhinoceros. Roughly 4 meters long from head to tail and about 2300 kilograms in mass, a large rhino is approximately the size of a small car. A rhino's enormous size, beady eyes, armored skin, curving horn (or horns), and stocky legs remind some people of the dinosaurs that roamed the Earth long ago.

Recently, it became apparent that rhinos were in great danger of sharing yet another characteristic with dinosaurs: being extinct. (The word extinct is used to describe a type of living thing that has died out completely and vanished forever.) Why? Because in the past 20 years, more than 90 percent of the world's rhinos have been killed for their horns.

African rhinos have two horns: a long one on the tip of their snout and a smaller one behind it. Asian rhinos have only one horn. In both types of rhinos, the horns are nothing more than large curved cones made of the same substance as your fingernails. But in some parts of the world, people believe that rhino horns have magical properties. In Yemen, for example, rhino horns are used to make the handles of special daggers. And in China, ground-up rhino horns are used in "medicines" for fevers and other ailments.

What can be done to stop the slaughter of the rhinos before it's too late? Groups of nations have agreed to prohibit trade in rhino horns. Individual countries have made rhino hunting illegal, set up protected wildlife parks, and supplied park rangers with better equipment for stopping illegal hunters. Zoos have developed ways of raising and breeding rhinos in captivity. And Michael Werikhe (WAIR-ree-kee), a young man from Kenya, has walked more than 7800 kilometers to make people aware of the threat to rhinos and to raise money for rhino conservation programs.

Michael Werikhe is not a politician, scientist, or professional fundraiser. He is a

142 ■ GAZETTE

security officer at an auto factory, and he doesn't particularly like to walk. But he is determined to do all he can to save rhinos from extinction.

As a boy, Werikhe played in the mangrove forests and explored the seashore near his home city of Mombasa. These childhood activities filled him with a deep appreciation for the living world. After high school, he took a government job that he thought might

TEACHING STRATEGY: ADVENTURE

FOCUS/MOTIVATION

Display pictures of the white and black rhinoceroses, elephants, condors, bald eagles, red kangaroos, gray wolves, greenback cutthroat trout, and leopards. Ask students what these animals have in common. (All are threatened or endangered species.)

CONTENT DEVELOPMENT

Explain that today poaching is a major reason for the plight of the rhinos.
- **What is poaching?** (Fishing or hunting illegally.)

The poachers hunt the rhinoceros for its horn. To them, the financial gain resulting from the sale of the horn is greater than the risk of being caught and punished. Stopping the poachers is not easy. They will continue the slaughter of

be related to his interest in wildlife. But it was not quite what he expected. He worked in a government warehouse that stored elephant tusks and rhinoceros horns confiscated from illegal hunters or collected from animals that had to be killed by game wardens. Werikhe's job was to sort these grisly trophies so that the government could sell them at auctions. Although Werikhe soon quit, the memory of the tusks, horns, and wasted wildlife they represented remained with him.

When Werikhe quit his job at the government warehouse, there were about 20,000 black rhinos in Kenya. Ten years later, there were about 500. To Werikhe, the rhinos became a symbol of all the threatened wildlife he loved. He had to do something. So he walked from Mombasa to Nairobi, a journey of about 480 kilometers. On his way, he stopped at many small villages and talked to people about the dangers rhinos faced.

"Too often, wildlife professionals assume that the average African cares little about preserving his natural heritage," Werikhe notes. "In my walks I've found that people *do* care, but feel left out."

In 1985, with the support of his employer and his fellow workers, Werikhe left his job and set off on a nearly 2100-kilometer trek across Africa. Three years later, he was off on another transcontinental hike. This time the continent was Europe. The walk raised $1 million for rhino conservation.

From April through September 1991, Werikhe, then 34 years old, undertook his third marathon walk. This walk took him to Washington, DC, Dallas, Toronto, San Diego, and many other major cities in North America. Three fourths of the approximately $3 million that he raised was for rhino programs in Africa; the rest was for rhino programs in North American zoos.

Werikhe has been honored for his work with a number of prestigious environmental

▲▼ In some parts of the world, people believe that rhino horns possess magical properties. As a result, African rhinos (top) and Indian rhinos (bottom) are in danger of becoming extinct.

awards, including the United Nations Environment Program's Global 500 Award. He has shown the world that the efforts of one ordinary person (with extraordinary determination) can go a long way toward solving global problems. But the accomplishment that might mean the most to Michael Werikhe is this: Thanks in large part to his efforts, the African rhino population seems to be back on the road to recovery.

GAZETTE ■ 143

Additional Questions and Topic Suggestions

1. Use references to find out about endangered species in the United States. Choose one endangered species and write about its natural habitat, behavior, and physical characteristics. Tell about the efforts that are being made to help the populations of the species grow.

2. Some environmental-awareness groups and animal-rights advocates may use tactics that dramatize their message but that may be considered radical or even illegal. Research the tactics and methods of operation of one of the groups, such as Greenpeace. Prepare a paper that supports or opposes the methods used by the group. Defend your position.

3. Compare and contrast the physical and behavioral characteristics of the African black rhino, the African white rhino, the Indian one-horned rhino, the Sumatran rhino, and the Java rhino.

4. Trace the evolution of the rhinoceros.

Critical Thinking Questions

1. Michael Werikhe devoted his personal efforts to help save the rhinoceros from extinction. Discuss in groups what you as individuals can do to help protect an endangered species. Plan to follow through on one of the suggestions.

2. Imagine that you are a reporter who has interviewed Michael Werikhe. Write a news article about the message Werikhe is trying to spread. In your article, include information about the plight of the rhinoceros and the efforts being made to help protect the animal.

3. Should there be legal trade of rhino horns obtained from rhinoceroses that died naturally, or should trade in rhino horns be banned completely?

rhinoceroses as long as there is a market for the horns.

As an individual, Michael Werikhe has raised funds that are used to protect the rhino, and he has helped to increase awareness of their plight. With the help of national and international groups that have also worked to help protect the rhino, Werikhe's efforts may save the rhino from extinction. In Werikhe's homeland, Kenya, the Rhino Rescue Trust was established. It has raised funds

and has set up a sanctuary for rhinos in Lake Nakuru National Park. Rhinoceroses are relocated in a sanctuary there, where they are protected by armed guards. Another group, Save the Rhino Trust, has been operating in Zambia since the early 1980s. These groups and others are working to save the rhinoceros in a variety of ways—education, sanctuaries, antipoaching efforts, and laws.

INDEPENDENT PRACTICE

▶ *Activity Book*

After students have read the Science Gazette article, you may want to hand out the reading skills worksheet based on the article in the *Activity Book*.

TUNA NET FISHING: DOLPHINS
IN DANGER!

Background Information

For unknown reasons, spotted dolphins, spinner dolphins, and common dolphins travel with schools of yellowfin tuna in the eastern Pacific. This fact has long been used by tuna fishers. Dolphins leaping out of the water signal the location of tuna. For thousands of years, people fished for tuna using rod, line, and baitless hook, none of which posed any threat to dolphins. That changed in the 1960s with the introduction of purse seines, huge nets that can be floated from the surface of the water and then drawn shut at the bottom, like the drawstring on a bag, trapping everything swimming below. It is estimated that between 250,000 and 500,000 dolphins were killed annually, with the total exceeding 6 million.

In the early 1970s, a public outcry against the dolphin slaughter led to a boycott of canned tuna and eventually to the passage of the Marine Mammal Protection Act of 1972, which called for reducing the dolphin kill each year until reaching "a zero mortality and serious injury rate." But the act was never completely implemented. Congress bowed to industry pressure, and over the years regulations were relaxed, enforcement was limited, and the allowable kill level was eventually frozen at the 1984 number: 20,500. And that kill number applied only to United States tuna boats; it did not include the number of dolphins killed by foreign boats. In January 1988, a coalition of environmental groups sponsored a tuna boycott against the three main United States importers of tuna. Public schoolchildren even demanded that tuna be taken off school cafeteria menus. In April 1990, the three companies announced that they would no longer accept tuna caught in the eastern Pacific unless it had been caught without harming dolphins.

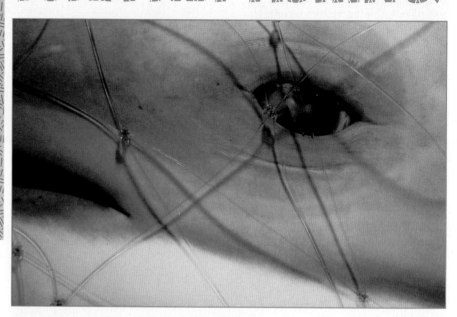

TUNA NET FISHING:

DOLPHINS IN DANGER!

Sorry, Charlie! Although tuna fish, whether in a sandwich, in a salad, or on a plate, remains a popular meal, some tuna-fishing methods are now considered unacceptable. Why is this so? Environmentalists have shown that purse seines, the very large rectangular fishing nets used to catch most tuna, have also caught and destroyed as many as 80,000 to 100,000 dolphins a year for the past several years. Fishing regulations permit a certain number of dolphins to be killed as a result of commercial fishing. However, the actual number of dolphin deaths is four to five times greater than that allowed by law.

Most tuna fishing takes place in the Pacific Ocean, where fishing boats from the United States as well as from the Republic of Korea, Taiwan, and Japan enjoy access to the fish-rich waters. There, fishers often find schools of yellow-fin tuna swimming with dolphins. In fact, fishers rely on the dolphins to help them locate large quantities of tuna. But the fishers' nets cannot tell the difference between fishes and mammals. The

TEACHING STRATEGY: ISSUE

FOCUS/MOTIVATION

Ask students to write a short description of a dolphin. Encourage them to include their ideas and feelings about dolphins, not simply facts that they know about dolphins.
• **Did you describe the dolphin as friendly, intelligent, or playful?** (Accept all answers.)

Ask students to name adjectives they used to describe the dolphin.
• **How do you picture the dolphin in your mind?** (Accept all answers. Students will probably picture the dolphin as nonthreatening or cute. Some may mention the old television show *Flipper* or other programs.)
• **How are the images that we have of dolphins likely to affect efforts to protect them?** (Accept all answers. Discuss the idea that people's fondness for dol-

▲▲ When dolphins are caught in nets (right), they cannot surface for air and they drown. Fortunately, this dolphin (left) was freed by a diver swimming nearby.

purse seines trap both tunas and dolphins under water. Unable to surface for their much-needed air, the dolphins drown.

Purse seines are not the only fishing equipment responsible for the deaths of vast numbers of marine life. Drift nets, which are huge fishing nets that placed together can stretch more than 4 kilometers across, have also produced criticism from scientists. Drift nets have been used to catch a wide variety of commercial species—including cod, tuna, and squid—since the late 1970s. But because of their size, drift nets ensnare and kill a tremendous number of other species of marine life, such as sharks, dolphins, sea turtles, and sea birds. In the last half of 1989, an estimated 1.5 million blue sharks, 23,000 dolphins, and 230,000 sea birds were killed in drift nets. The nets may also be responsible for reducing some of the commercial fish supply beyond a healthy number, thus interfering with the development of young fishes. So although scientists still do not have all the data they need to completely assess the situation, they are expressing alarm at the threat the drift nets may pose to marine ecosystems.

Scientists are not the only people concerned with purse seines and drift nets. The general public as well as national and international governing bodies have been part of a global effort to protect dolphins—as well as other types of marine life—from the nets. For instance, various concerned people—including schoolchildren—used boycotts and crusades to convince three major American tuna marketers to stop buying tuna from fishing companies that used purse seines. Tuna marketers now put special labels on their cans to show that the tuna inside was caught by relatively "dolphin-safe" methods.

In 1990, the United Nations passed a resolution that would prohibit drift-net fishing in the South Pacific by 1991 and worldwide by 1992. The resolution, however, leaves room for drift-netters: They will be allowed to use their nets if they take "effective conservation and management measures." These measures include releasing the marine life that accidentally gets trapped in the nets.

GAZETTE ■ 145

1. People have been drawn to dolphins for thousands of years. Dolphins are perceived as attractive, charming, smiling creatures. In addition to their strong emotional appeal, dolphins are considered to be highly intelligent and non-threatening, even friendly, to humans. Do you think there would have been such a loud public outcry if dolphins were ugly, vicious, and dangerous to humans?

2. Investigate the passage and subsequent history of the Marine Mammal Protection Act (1972). What were its original provisions? How many of those provisions were implemented? What happened to the act over the next two decades? How does the fate of this act show that the passage of legislation does not necessarily mean that a problem has been solved?

3. Inevitably, conservation issues are also economic issues. Restrictions on tuna fishing could lead to job losses and higher prices for consumers. And that could happen in many other industries. Consider the recent controversy that has pitted the spotted owl against the loggers in the Pacific Northwest. How can you convince people that the survival of a species is worth the loss of their jobs?

---

phins will probably make them more interested in helping the animals.)

CONTENT DEVELOPMENT

Point out that purse seines and drift nets both came into use at a time when commercial fishers were under pressure to increase their catches in order to compete. Larger nets mean larger and quicker catches. Explain that the traditional rod, line, and baitless hook method is not seen as economically viable for a large fishing company.

• **Why is the use of purse seines and drift nets economical for fishers and dangerous to dolphins and other marine life?** (The nets allow fishers to catch huge quantities of fish at one time, but the nets also catch and kill dolphins and other sea creatures by drowning, strangling, or crushing them.)

Explain that as with many global environmental problems, international cooperation is necessary to protect dolphins and other marine life. The United

Nations resolution on drift-net fishing is an important first step because it is based on the agreement and involvement of many nations.

• **What is the danger in allowing fishers to continue to use drift nets?** (Even though fishers are supposed to release marine life that gets trapped in their drift nets, they may simply not do it, or they may not be able to save all the trapped animals.)

Class Debate

Point out that stopping the killing of dolphins can be seen as part of a wider movement to protect animals from all forms of abuse. Divide the class into teams to debate the following issue:

Animals should not be used in any types of tests or experiments, regardless of how urgent or important these tests or experiments might be to humans.

▲ Leaping dolphins have captivated people's imagination for centuries.

But some environmental groups are still worried. They fear that because the nets catch such a wide variety of marine species, it may be impossible to take the effective measures necessary to protect all the species.

Meanwhile, other groups are responding to the fishing restrictions in other ways. Members of the tuna-fishing industry argue that companies refusing to do business with fishers who use purse seines may actually end up doing more harm than good. They explain that fishers will be forced to catch mostly the young tuna that do not swim with dolphins. This action would ultimately reduce the supply of tuna. And they add that the cruelty of their industry is highly exaggerated. In fact, they claim that 90 percent of the tuna caught in the Pacific

Ocean is harvested in ways that provide a minimal threat to dolphins.

Officials from various countries also fear for the economic safety of their businesses and workers. Safer nets will not only cost more money, they will also increase competition by limiting the availability of fishing sites. And in the end, it will be the consumers who will feel these effects by having to pay a higher price for the tuna.

Fishing practices and the responsibilities of people to protect water wildlife have sparked debate in homes, classrooms, and national and international congresses. What do you think?

▼ The label "dolphin safe" indicates that the tuna inside this can was not caught in nets that also kill dolphins.

146 ■ GAZETTE

Issue (continued)

GUIDED PRACTICE

Skills Development

Skill: Investigating an issue

Have students answers the question at the end of the article. You may wish to use the question as a basis for a class discussion or have students work in teams to debate the issue. Another possibility would be to have small groups of students write a set of questions about the issue and conduct a survey among their friends and relatives. They could present their findings in an oral report or in chart form. Some students might want to research the threat to the fishing industry in more detail. Others might like to think of ways to raise public awareness of the dangers of drift-net fishing by creating posters, ads, slogans, or billboards.

REINFORCEMENT/RETEACHING

Encourage students to organize the information in the article in a way that will show them how the statements are related. Ask them to write a series of cause-and-effect questions. Each question should be answered with a statement that begins with *because*. For example, students might begin with this:

Why do fishers look for dolphins? Because dolphins swim with tuna, and the fishers use the dolphins to help them find the tuna.

Why are the fishers' nets dangerous to dolphins? Because the nets catch the dolphins and hold them under water so that they drown.

Have pairs of students take turns asking and answering each other's series of questions.

INDEPENDENT PRACTICE

▶ *Activity Book*

After students have read the Science Gazette article, you may want to hand out the reading skills worksheet based on the article in the *Activity Book*.

The *HIT* of the Class of 2025

THE *HIT* OF THE CLASS OF 2025

The wallscreen glowed faintly as George sat quietly at his television console. His heart pounded faster and faster as he heard the announcer's voice. "T minus five minutes and counting."

The picture on the screen showed the largest spaceship ever built on Earth. Inside, a team of astronauts was prepared for

Hidden in the blood of hibernating animals is a newly discovered chemical. It could slow the aging process of humans and make years-long journeys into space possible.

Background Information

Hibernation is a time during which certain animals sleep through the winter in caves or burrows. Among animals that hibernate are woodchucks, squirrels, hamsters, and bears.

Although people have been fascinated by hibernation for centuries, it is only since the 1960s that scientists have begun to understand it. Prior to that time, scientists assumed that hibernation in certain animals was a natural response triggered by the onset of cold weather. In 1968, however, two American physiologists, Wilma Spurrier and Albert Dawe, proposed the hypothesis that hibernation is triggered by a specific chemical.

Spurrier and Dawe began to test their hypothesis by extracting blood from a hibernating squirrel that was asleep in a cold laboratory designed to simulate winter. They then injected the blood into two active squirrels. Within 48 hours, the two squirrels were hibernating. The following summer, the two physiologists injected blood saved from the previous winter from hibernating squirrels into 23 active squirrels. Twenty of the 23 squirrels began to hibernate in the middle of summer!

By now, the researchers were convinced that something in the blood of a hibernating animal triggers hibernation. They named the chemical HIT, which stands for hibernation-induction trigger. Still to be accomplished was the isolation of HIT from the blood. A biochemist named Peter Oeltgen took up the project at this point and determined that HIT is contained in the albumin. At the present time, Oeltgen and his colleagues are seeking to discover the amino-acid sequence of HIT.

TEACHING STRATEGY: FUTURE

FOCUS/MOTIVATION

Display pictures of squirrels, hamsters, woodchucks, and bears.
• **What do these animals all have in common?** (Accept all answers.)

CONTENT DEVELOPMENT

Continue the previous discussion by pointing out that all the animals pictured are hibernators; that is, they enter into a deep sleep that carries them through the winter. When spring comes, usually around March, the animals wake up and become active again.

Additional Questions and Topic Suggestions

1. Get together with several classmates and create a play about people who hibernate for a period of time and then wake up. Present your play to the rest of the class.

2. Can you think of any possible dangers or risks involved in human hibernation? (Accept all answers. One danger could be that different people might have different physiological reactions to HIT; if body functions were to slow down too much, for example, a person could die or suffer brain damage.)

3. Use reference sources to learn about the hibernation patterns of various animals. Create a classroom display in which you include this information with a picture of each animal.

▲ Ground squirrels enter a state of hibernation during the cold winter months.

Future (continued)

CONTENT DEVELOPMENT

Point out that hibernation is not ordinary sleep. In fact, in the 1700s a naturalist found a bird that was hibernating and thought the bird was dead, not sleeping. During hibernation, many of the animal's body functions slow down. The heart rate decreases, often to half or less of its normal rate. Brain activity is reduced, and the endocrine system is suppressed. Respiration slows down and oxygen consumption decreases. The animal's body temperature falls approximately 30 degrees C. During hibernation, animals live on food that is stored in the body as fat. That is why some animals look so fat in the fall—they are getting ready for their long winter's nap.

• **How do you think animals benefit from hibernation?** (Answers may vary. Probably the greatest benefit is that the animals do not have to search for food in the winter, when food is hard to find. Also, they are able to stay warm without having to brave severe winter weather.)

the longest space journey ever attempted by humans—a five-year trip to the outer planets and their moons. The astronauts planned to "sleep" through most of the trip as automatic devices guided the spaceship to its destination. But how could humans sleep for days, weeks, even months without eating or drinking?

The answer, which George knew, was simple. The astronauts would hibernate! During hibernation, their body temperatures would be lowered. All their bodily activities would slow down considerably, which is exactly what happens to the bodies of hibernating animals such as squirrels and dormice.

But, you might think, people do not hibernate. Well, not now, but by the year 2025 . . .

GETTING HIT MAKES SENSE

"As you know," the announcer said, "the astronauts will soon inject themselves with a substance called hibernation-induction trigger, or HIT for short. This substance was discovered in the blood of hibernating animals back in the 1970s. Perhaps more startling, when HIT first was injected into nonhibernating animals, the animals promptly went to sleep. Now that HIT has been developed for use in humans, astronauts can be sent on spaceflights that take months or years. They'll just sleep through the flights. They won't get bored or need food."

George thought about other ways in which HIT could be used. For example, scientists had discovered about 40 years earlier that cancer does not spread in hibernating animals. This means that cancer patients could be put into a state of hibernation while receiving special treatments. Also, a hibernating body can take much more stress than an active body can. So patients could be placed in hibernation during surgery. This would eliminate the risks involved in using drugs to put patients to sleep. In addition, because bodies age very little while hibernating, people's lives could be extended considerably if they hibernated from time to time.

This last thought threw George a bit. "I don't know if I'd want to live longer if I had to sleep away every winter like a squirrel," he thought to himself. "I'd miss skiing and ice skating and . . . " George's thoughts were interrupted by his older sister's voice.

Continue the discussion by explaining that until 1968, scientists did not really understand what causes an animal to hibernate. Then, two American physiologists proposed the idea that a certain chemical induces hibernation. By injecting blood from hibernating animals into active animals, the physiologists proved that this idea was correct. That is how HIT was discovered.

CONTENT DEVELOPMENT

Discuss the possible use of HIT to prevent aging. Explain that the study of aging is a relatively new branch of science and that researchers are just beginning to understand what causes some organisms to age more rapidly than others.

Students may be interested to know that the creatures on the earth today that live the longest are tortoises. The oldest

"It's two o'clock, George. Aren't you going to the senior barbecue?"

"I almost forgot!" George exclaimed as he leaped out of his chair and headed for the door.

"How could you forget the farewell party for the Class of 2025?" his sister teased as George dashed out of the house.

Behind George, the wallscreen still glowed. Suddenly, a huge flaming cloud of gas burst from the bottom of the rocket shown on the screen. Gaining speed with each passing second, the rocket rose faster and faster, launching its crew on the first leg of its five-year mission. The astronauts would return feeling and looking only about one year older! HIT would help keep them young.

STAYING YOUNGER

As George arrived at the barbecue, his mind wandered back to the scene on the wallscreen and the announcer's description of HIT. Again his thoughts were interrupted by a voice.

"Here's to the reunion of 2050," called one of George's friends.

"And to the one in 2075," cried another friend. The class had decided that no matter how far its members would wander, they would all get together every 25 years.

"Ugh," groaned still another friend, "we'll all be old, wrinkled, and gray by then."

"Maybe not," George said with a smile.

"What are you talking about?" asked a number of friends at the same time.

George paused for a moment in order to build up the suspense.

"What if we were able to hibernate for eight hours a day instead of sleeping for the same amount of time?"

"We'd turn into squir-

rels," joked someone, causing everyone to laugh.

"Not quite," said George. "We just might not turn into old people as fast as we would normally."

"I don't get it," stated a classmate.

"Simple," said George. "Eight hours is a third of a day. If we didn't age—or, at least didn't age much—for a third of the time, we'd only age about 18 years out of every 25."

"You mean in 2050 I'd look like 36 instead of 43?" came a question from the crowd.

"Maybe. And in 2075 you'd look like you were in your early fifties instead of your late sixties."

"Is it really possible?" someone asked.

A blur of gray fur caught George's eye as he was about to answer. The blur raced up the side of a nearby tree. George stared at the little animal with the bushy tail.

"You might ask the squirrel," suggested George. "He knows the secret."

▼ **These people—all of whom are at least 100 years old—live in certain areas in Russia. Scientists wonder why.**

GAZETTE ■ 149

Critical Thinking Questions

1. If you were going to hibernate, which part of a year would you prefer to miss? Which years of the next several decades might you prefer to sleep through? (Accept all answers.)

2. Can you think of possible reentry problems for people who hibernate for months or years? (Accept all answers. Possible answers include psychological problems caused by reentering a world or immediate environment that has changed during the time one was asleep.)

3. Can you think of practical problems for society and the economy if hibernation became commonplace among people? (Accept all answers. Possible answers include the problem of unpaid bills while a person is asleep; the problem of a person being absent from his or her job; difficulty collecting income taxes, and so on.)

tortoise is about 170 years old, and it is not uncommon for tortoises to live to be well over 100. Second to tortoises in longevity are humans. Scientists believe that the oldest human is a man in Japan who is 118.

Scientists attribute the long life of tortoises to their slow pace of life, which reduces the organism's demand for energy. Humans, on the other hand, survive as

long as they do because of the body's ability to repair cell damage efficiently.

INDEPENDENT PRACTICE

▶ *Activity Book*

After students have read the Science Gazette article, you may want to hand out the reading skills worksheet based on the article in the *Activity Book*.

or Further Reading

If you have been intrigued by the concepts examined in this textbook, you may also be interested in the ways fellow thinkers—novelists, poets, essayists, as well as scientists—have imaginatively explored the same ideas.

Chapter 1: Sponges, Cnidarians, Worms, and Mollusks

Headstrom, Richard. *Suburban Wildlife*. Englewood Cliffs, NJ: Prentice Hall.

Henry, O. "The Octopus Marooned." In *The Complete Works of O. Henry*. New York: Doubleday.

Rockwell, Thomas. *How to Eat Fried Worms*. New York: Franklin Watts.

Sroda, George. *Life Story of TV Star and Celebrity Herman the Worm*. Amherst Junction, WI: G. Sroda.

Terris, Susan. *Octopus Pie*. New York: Farrar, Straus & Giroux.

Chapter 2: Arthropods and Echinoderms

Fleischman, Paul. *Joyful Noise: Poems for Two Voices*. New York: Harper & Row.

Goor, Ron, and Nancy Goor. *Insect Metamorphosis: From Egg to Adult*. New York: Atheneum.

Herberman, Ethan. *The Great Butterfly Hunt: The Mystery of the Migrating Monarchs*. New York: Simon & Schuster.

Kingsley, Charles. *The Water-babies*. New York: Dodd, Mead.

Kipling, Rudyard. *Just So Stories*. New York: Harper & Row.

Webster, Jean. *Daddy Long-Legs*. New York: Scholastic.

White, E. B. *Charlotte's Web*. New York: Harper & Row Junior Books.

Chapter 3: Fishes and Amphibians

Cole, Harold. *A Few Thoughts on Trout*. New York: Messner.

Dobrin, Arnold. *Going to Moscow and Other Stories*. New York: Four Winds.

East, Ben. *Trapped in Devil's Hole*. Riverside, NY: Crestwood.

George, Jean Craighead. *Shark Beneath the Reef*. New York: Harper & Row.

Kipling, Rudyard. *Captains Courageous*. New York: Airmont.

Chapter 4: Reptiles and Birds

Aesop. *Aesop's Fables*. New York: Avenel Books.

Blotnick, Elihu. *Blue Turtle Moon Queen*. Canyon, CA: California Street Press.

Edler, Timothy J. *Maurice the Snake and Gaston the Nearsighted Turtle*. Loreauville, LA: Little Cajun Books.

Fleischman, Paul. *I Am Phoenix: Poems for Two Voices*. New York: Harper & Row.

Henry, O. "A Bird of Baghdad." In *The Complete Works of O. Henry*. New York: Doubleday.

Lavies, Bianca. *The Secretive Timber Rattlesnake*. New York: Dutton.

Yolen, Jane. *Bird Watch: A Book of Poetry*. New York: Philomel.

Chapter 5: Mammals

Bowman, James Cloyd, and Margery Bianco. *Tales from a Finnish Tupa*. Chicago, IL: Albert Whitman.

Burnford, Sheila. *The Incredible Journey*. Boston, MA: Little, Brown.

George, Jean Craighead. *Julie of the Wolves*. New York: Harper & Row.

Kipling, Rudyard. *The Jungle Book*. New York: New American Library.

Nash, Ogden. *The Animal Garden*. New York: M. Evans.

Patent, Dorothy Hinshaw. *Seals, Sea Lions, and Walruses*. New York: Holiday House.

Rawlings, Marjorie Kinnan. *The Yearling*. New York: Macmillan.

Activity Bank

Welcome to the Activity Bank! This is an exciting and enjoyable part of your science textbook. By using the Activity Bank you will have the chance to make a variety of interesting and different observations about science. The best thing about the Activity Bank is that you and your classmates will become the detectives, and as with any investigation you will have to sort through information to find the truth. There will be many twists and turns along the way, some surprises and disappointments too. So always remember to keep an open mind, ask lots of questions, and have fun learning about science.

Chapter 1   SPONGES, CNIDARIANS, WORMS, AND MOLLUSKS

TO CLASSIFY OR NOT TO CLASSIFY?   **152**

FRIENDS OR FOES?   **153**

MOVING AT A SNAIL'S PACE   **154**

Chapter 2   ARTHROPODS AND ECHINODERMS

OFF AND RUNNING   **156**

SPINNING WEBS   **157**

HOW MANY ARE TOO MANY?   **159**

Chapter 3   FISHES AND AMPHIBIANS

TO FLOAT OR NOT TO FLOAT?   **161**

Chapter 4   REPTILES AND BIRDS

DO OIL AND WATER MIX?   **163**

AN EGGS-AGGERATION   **164**

STRICTLY FOR THE BIRDS   **165**

Activity Bank

COOPERATIVE LEARNING

Hands-on science activities, such as the ones in the Activity Bank, lend themselves well to cooperative learning techniques. The first step in setting up activities for cooperative learning is to divide the class into small groups of about 4 to 6 students. Next, assign roles to each member of the group. Possible roles include Principal Investigator, Materials Manager, Recorder/Reporter, Maintenance Director. The Principal Investigator directs all operations associated with the group activity, including checking the assignment, giving instructions to the group, making sure that the proper procedure is being followed, performing or delegating the steps of the activity, and asking questions of the teacher on behalf of the group. The Materials Manager obtains and dispenses all materials and equipment and is the only member of the group allowed to move around the classroom without special permission during the activity. The Recorder, or Reporter, collects information, certifies and records results, and reports results to the class. The Maintenance Director is responsible for cleanup and has the authority to assign other members of the group to assist. The Maintenance Director is also in charge of group safety.

For more information about specific roles and cooperative learning in general, refer to the article "Cooperative Learning and Science—The Perfect Match" on pages 70–75 in the *Teacher's Desk Reference*.

ESL/LEP STRATEGY

Activities such as the ones in the Activity Bank can be extremely helpful in teaching science concepts to LEP students—the direct observation of scientific phenomena and the deliberate manipulation of variables can transcend language barriers.

Some strategies for helping LEP students as they develop their English-language skills are listed below. Your school's English-to-Speakers-of-Other-Languages (ESOL) teacher will probably be able to make other concrete suggestions to fit the specific needs of the LEP students in your classroom.

• Assign a "buddy" who is proficient in English to each LEP student. The buddy need not be able to speak the LEP student's native language, but such ability can be helpful. (**Note:** *Instruct multilingual buddies to use the native language only when necessary, such as defining difficult terms or concepts. Students learn English, as all other languages, by using it.*) The buddy's job is to provide encouragement and assistance to the LEP student. Select buddies on the basis of personality as well as proficiency in science and English. If possible, match buddies and LEP students so that the LEP students can help their buddies in another academic area, such as math.

• If possible, do not put LEP students of the same nationality in a cooperative learning group.

• Have artistic students draw diagrams of each step of an activity for the LEP students.

You can read more about teaching science to LEP students in the article "Creating a Positive Learning Environment for Students with Limited English Proficiency," which is found on pages 86–87 in the *Teacher's Desk Reference*.

Activity Bank

TO CLASSIFY OR NOT TO CLASSIFY?

BEFORE THE ACTIVITY

1. Divide the class into groups of two students. If you wish, you can increase the number of students in each group to four.

2. Gather the index cards and 20 photographs of plants at least one day prior to the activity. You should have enough materials to meet your class needs.

PRE-ACTIVITY DISCUSSION

Have students read the What Do You Need to Do? (procedure) in both parts of the activity. Discuss the What Do You Need to Do? by asking questions similar to the following.

• **What is the purpose of this activity?** (To study the need for classification.)

• **What are some characteristics that can be used to divide and subdivide the letters of the alphabet and the plants?** (Answers will vary, but lead students to discuss appropriate characteristics. A brief list of appropriate characteristics might be placed on the chalkboard.)

TEACHING STRATEGY

1. As groups develop their classification systems, have students write out a branching chart with the largest group(s) at the top and the smallest subgroups at the bottom. Have one member of each group prepare a chart for the entire group.

2. At the conclusion of the activity, have the group members who prepared the chart for each group copy their group's branching chart on the chalkboard.

DISCOVERY STRATEGIES

Discuss how the activity relates to the ideas presented in the chapter by asking questions similar to the following.

• **You learned that all living things are classified into five major groups—monerans, protists, fungi, plants, and animals. How is the classification system that you developed similar to these groups?** (Answers will vary—making comparisons, classifying.)

TO CLASSIFY OR NOT TO CLASSIFY?

Have you ever tried to organize the clothes in your room? If so, perhaps you put all your sweaters into one group, your socks into another group, and your jeans into still another group. In other words, you classified your clothes into specific groups based on their common characteristics. Why is it useful to classify objects? Try this activity and find out.

Part A. You will need 26 index cards and a pencil for this activity.

What Do You Need to Do?

1. Write each letter of the alphabet on an index card. Use only capital letters.

2. Choose any trait that will enable you to classify the letters. Arrange the letters in their appropriate groups.

3. After you have classified the letters according to one trait, choose another trait and again classify the letters into groups.

Index cards

What Did You Find Out?

1. What traits did you use to classify the letters of the alphabet?

2. How were the letters arranged in each group?

Part B. Now try your hand at classifying living things. You will need 20 photographs of plants for this activity.

What Do You Need to Do?

1. Collect 20 photographs of plants.

2. Classify the plants according to color.

3. Then classify the plants according to whether they live in water or on land.

4. Examine the photographs of the plants again. Determine whether another trait will help you classify the plants. Arrange the plants in their appropriate groups.

5. After you have classified the plants into their specific groups, name each group based on the trait that best describes it.

6. Give your photographs of the plants to a classmate and tell your classmate the names you have given each group. Have your classmate use your group names to classify the photographs. Then switch roles and try to do the same with your classmate's group names.

What Did You Find Out?

1. Is color a good way to classify plants? Explain your answer.

2. Is where a plant lives a good trait to use in classifying plants?

3. Did you and your classmate match each other's photographs with the appropriate group names?

4. Did either of you have any difficulty in doing so? If either of you did, what changes could be made in order to match the group name with the appropriate photograph?

5. Can you think of other traits that may be better for classifying plants?

6. Why do you think it is helpful to classify objects?

WHAT DID YOU FIND OUT?: Part A

1. Answers will vary. Students may place the letters in groups that contain letters with only straight lines, and those with curves.

2. Answers will vary.

WHAT DID YOU FIND OUT?: Part B

1. Students should realize that color is not a good way to classify plants because most plants are green. They, therefore, would all be placed in the same group.

2. For the most part, plants live on land. Classifying them in this manner would result in placing them into one large group.

3. Answers will vary, depending on whether a student's classification system was logical and clearly named.

4. Answers will vary. Students should give logical explanations as to what types of changes can be made to improve a classification system.

5. Student answers may include: flowering and nonflowering plants, woody and non-woody plants, shape of leaves, and so on.

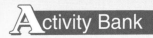
FRIENDS OR FOES?

Do animals and plants need each other? Can they survive without one another? Try this activity to find out the answers to these questions.

What Will You Need?

small plastic bag with a twist tie
distilled water
medicine dropper
bromthymol blue solution
pond snail
Elodea sprig

What Will You Do?

1. Fill a plastic bag two-thirds full with distilled water.
2. Put a few drops of bromthymol blue solution into the plastic bag to make the water blue.
3. Put a pond snail into the plastic bag. Seal the bag with a twist tie and put it in a place where it will remain undisturbed for 20 minutes.

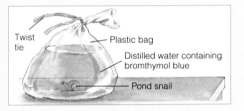

4. After 20 minutes, observe the color of the water. Bromthymol blue solution will turn yellow in the presence of carbon dioxide.
5. Unseal the plastic bag and add an *Elodea* sprig. Reseal the plastic bag and place it in an area that receives light (not direct sunlight) for 20 minutes.

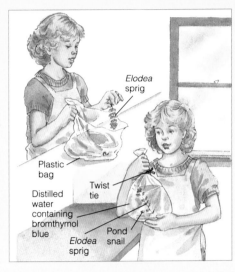

6. After 20 minutes, observe any changes in the color of the water.

What Did You See?

1. What was the color of the water containing the pond snail after 20 minutes?
2. What was the color of the water containing the pond snail and *Elodea* sprig after 20 minutes?

What Did You Discover?

1. What happened to the carbon dioxide gas that was produced by the snail?
2. Do animals and plants need each other? Explain.
3. Many trees in the Amazon rain forest are being destroyed to clear the land for farming and ranching. Explain how the disappearance of the rain forest would affect the gases in the environment.

C ■ 153

• **What is the purpose of the activity?** (To show how plants and animals are dependent on one another; to observe the interactions of plants and animals.)
• **Do plants need oxygen? Do animals?** (Plants and animals both need oxygen for respiration.)
• **Do plants and animals need carbon dioxide?** (Plants need carbon dioxide for photosynthesis; animals do not need carbon dioxide.)

TEACHING STRATEGY

1. If you do not have sufficient time to do this activity in one period, you may wish to do steps 1 through 4 on one day and steps 5 and 6 on the next day. Make sure that the bromthymol blue has changed color from blue to yellow before you have students perform steps 5 and 6.
2. You might have to adjust the number of snails and the amount of *Elodea* to produce the desired results.

DISCOVERY STRATEGIES

Discuss how the activity relates to the chapter ideas by asking questions similar to the following.
• **Why do plants and animals need each other?** (Plants and animals provide each other with the necessary raw materials for respiration and photosynthesis. During respiration, plants and animals take in oxygen and give off carbon dioxide. During photosynthesis, plants take in carbon dioxide and give off oxygen—applying concepts.)

WHAT DID YOU SEE?

1. Yellow.
2. Blue.

WHAT DID YOU DISCOVER?

1. The carbon dioxide was used by the plant during the process of photosynthesis to produce oxygen.
2. Yes; during respiration, animals (and plants) produce carbon dioxide, which is needed by plants during photosynthesis. During this process, plants release oxygen, which is needed by animals (and plants).
3. During photosynthesis, plants take in carbon dioxide and release oxygen. If the rain forest disappears, the atmosphere would contain more carbon dioxide and less oxygen.

Activity Bank

FRIENDS OR FOES?

BEFORE THE ACTIVITY

1. Gather all materials at least one day prior to the activity. Distilled water can be purchased in a large supermarket. You can obtain pond snails from a biological supply company or gather them from a nearby pond. Make sure students treat the pond snails with care. Return the pond snails to their natural habitat.
2. Bromthymol blue is a pH indicator. The pH value at which bromthymol blue undergoes a color change from yellow to blue is 6.0–7.5.
3. If *Elodea* is not available, other aquatic plants such as hornwort (*Ceratophyllum*) can be used.

PRE-ACTIVITY DISCUSSION

Have students read the complete What Will You Do? (procedure).

Activity Bank

MOVING AT A SNAIL'S PACE

BEFORE THE ACTIVITY

1. Gather all materials at least one day prior to the activity. You should have enough supplies to meet your class needs.
2. Pond snails may be ordered from a biological supply company. Make sure that pond snails are small enough to fit on microscope slides.

PRE-ACTIVITY DISCUSSION

Have students read the complete procedure for the activity.
• **What is the purpose of this activity?** (To study what type of food snails prefer and to study how fast snails move.)
• **What kind of food do snails like to eat?** (Accept all hypotheses at this point.)
• **How do snails move? Do you think they can move very quickly?** (Accept all hypotheses at this point.)
• **What is the purpose of covering the microscope slide with sandpaper?** (To see if a different type of surface has an effect on the rate at which snails move.)

SAFETY TIPS

1. Remind students that pond snails, like all living organisms, must be treated with care and in a humane manner.
2. Microscope slides should be checked before use to make sure there are no rough or jagged edges. Caution students to use care in handling microscope slides.

TEACHING STRATEGY

1. At the beginning of the activity, distribute the snails to students.
2. Circulate about the room as students work to ensure they stay on task.
3. Make sure students handle the snails with care and do not keep them out of the pond water for extended amounts of time.

DISCOVERY STRATEGIES

Discuss how the activity relates to the chapter ideas by asking questions similar to the following.
• **To which group of animals does the snail belong?** (Mollusks—observing, making comparisons, applying facts.)
• **What characteristics make the snail a member of this group of animals?** (Snails are soft-bodied invertebrates that generally have inner or outer shells. These features are characteristic of mollusks—observing, making comparisons, relating.)
• **Is the snail an invertebrate or a vertebrate? How do you know?** (The snail is an invertebrate because it does not have a backbone—observing, making comparisons, applying facts.)

After insects, mollusks are the most varied group of animals on the Earth. Most mollusks have bodies that are covered by a shell. This activity will help you become more familiar with one small member of this group—a pond snail.

Materials

large dish	forceps
pond water	paper towel
pond snail	microscope slide
scissors	clock with second
metric ruler	indicator
lettuce and	2 cm x 6 cm piece
spinach leaves	of sandpaper

Procedure 🧪 🐌

1. Half fill the large dish with pond water and place the pond snail in the dish.
2. With the scissors, cut a 1-cm^2 section from the lettuce leaf and from the spinach leaf. Put the leaf squares in the dish with the pond snail.
3. Place the dish containing the pond snail and leaf squares in a place where they will remain undisturbed for 24 hours.
4. After 24 hours, remove the leaf squares from the dish and place them on the paper towel to remove excess water. Note how much of each square has been eaten.
5. Arrange the microscope slide in the dish so that the slide leans at a 45° angle against the side of the dish.
6. Place the snail at the top of the slide.
7. Measure the time it takes the snail to travel down and then up the microscope slide. Record the time for each.
8. Cover the microscope slide with the piece of sandpaper and repeat steps 5 through 7.

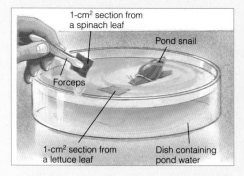

1-cm^2 section from a spinach leaf
Pond snail
Forceps
1-cm^2 section from a lettuce leaf
Dish containing pond water

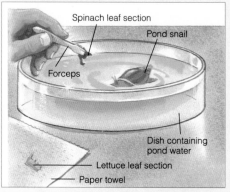

Spinach leaf section
Pond snail
Forceps
Dish containing pond water
Lettuce leaf section
Paper towel

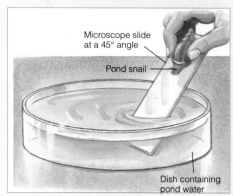

Microscope slide at a 45° angle
Pond snail
Dish containing pond water

OBSERVATIONS

1. Answers will vary. The snail is a herbivore because it eats plants.
2. Answers will vary.
3. Answers will vary.

ANALYSIS AND CONCLUSIONS

1. Answers will vary. The food that has the least remaining is probably the snail's favorite.
2. Snails travel faster going down the slide than they do going up the slide. This is because it takes less effort to travel down a

Observations

1. How much of each food was eaten? Is the snail a carnivore or a herbivore?
2. How long did it take the pond snail to travel down the microscope slide? To travel up the slide?
3. How long did it take the pond snail to travel down the microscope slide covered with sandpaper? To travel up the slide?

Analysis and Conclusions

1. What is the pond snail's favorite food? How were you able to determine this?

2. Do snails travel up and down at the same rate? If not, what do you think accounts for the difference?
3. Share your results with your classmates. Were the results similar? Were they different? If they were different, explain why.
4. How does the texture of a surface affect the pond snail's speed?

Going Further

You may wish to repeat steps 5 through 7 in the activity using other materials to cover the microscope slide. Determine how fast the snail can travel over the surfaces of these materials.

slanted surface than it does to travel up that same surface.
3. Student results should be similar.
4. The more textured the surface, the slower the snail's speed. This is because there is less friction (a force that brings an object to rest) when the snail moves across the smooth microscope slide than when the snail moves across the slide covered with sandpaper.

GOING FURTHER: ENRICHMENT

Have students try the experiment they designed in Going Further. If time does not permit this to be completed in class, interested students could perform their experiments as out-of-class projects.

Activity Bank

BEFORE THE ACTIVITY

1. Decide if students are to work in pairs or groups.
2. Gather all materials one day prior to performing the activity.
3. Pill bugs may be purchased from a biological supply company. The genus name for pill bugs is *Armadillidium*.

SAFETY TIPS

Remind students that pill bugs, like all living organisms, must be treated with care and in a humane manner.

TEACHING STRATEGY

1. At the beginning of the activity, distribute the pill bugs to students.
2. Circulate about the room as students work to ensure they stay on task.
3. Make sure students handle the pill bugs with care and do not handle them for extended amounts of time. Students may have to wait awhile until the pill bugs uncurl from their tight, round pill shapes before they begin to move.
4. Have students return the pill bugs to you so that you can place them in their natural habitat.

DISCOVERY STRATEGIES

Discuss how the activity relates to the chapter ideas by asking questions similar to the following.
• **Is the pill bug an insect?** (No. Insects have six legs and a body with three sections; pill bugs have 14 legs and a body with several sections—observing, making comparisons.)
• **How is the pill bug like an arthropod?** (The pill bug has an exoskeleton, a segmented body, and jointed appendages—observing, applying facts.)

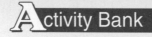
Sow bugs are also called pill bugs because when disturbed they curl up into a tight, round pill shape. Pill bugs, which are really crustaceans and not insects, hibernate in the winter. However, they can be found under rocks or under piles of dead leaves close to the foundation of a house. To find out more about pill bugs, hold a pill bug derby. All you will need is a pencil, a sheet of unlined white paper, a pill bug, a clock with a second indicator, and a metric ruler.

What Do You Need to Do?

Have each member of the class use a pencil to mark an X in the middle of a sheet of unlined white paper. One member of your class should act as the official timer. When the timer shouts "Go!" each person should place their pill bug on the X and let it go. The timer should clock one minute while you track your racer with a pencil line. When the minute is up, measure the distance of your racer's trail. The pill bug that travels the farthest is the winner. On your mark, get ready, go!

What Did You Discover?

1. How many legs do pill bugs have?
2. Describe the other characteristics of pill bugs.
3. Based on the above information, to what group of animals do pill bugs belong?
4. Share with your classmates the distance your pill bug traveled. What was the greatest distance a pill bug traveled? The least distance?

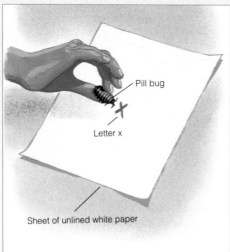

Pill bug

Letter x

Sheet of unlined white paper

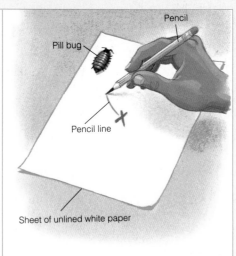

Pencil

Pill bug

Pencil line

Sheet of unlined white paper

WHAT DID YOU DISCOVER?

1. 14 legs.
2. Pill bugs have segmented bodies with a pair of legs attached to each segment.
3. Pill bugs are members of the phylum *Arthropoda*; more specifically, they are crustaceans.
4. Answers will vary.

SPINNING WEBS

Do all spiders produce silk? Yes, they do. Some spiders, however, do not use the silk to build webs. Those spiders that do use silk to build webs weave intricate patterns. To see how a spider goes about spinning its web, or nest, perform this activity.

What You Will Need

wire coat hanger
wood block (10 cm x 40 cm x 1 cm)
large, clear plastic bag with twist tie
black-and-yellow garden spider

What You Will Do

1. Remove the hooked part of a wire coat hanger by bending it back and forth a few times.

2. Bend the coat hanger into the shape of a square. Then insert the hanger, sharp side down, into a wood block so that it is supported by the wood block as shown in the drawing. This is called a net frame.

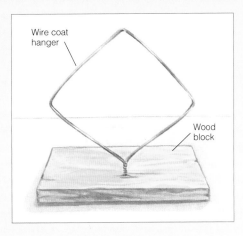

3. Blow up a large plastic bag and put the net frame inside.

4. Quickly place a black-and-yellow garden spider on the wood block and seal the plastic bag with the twist tie.

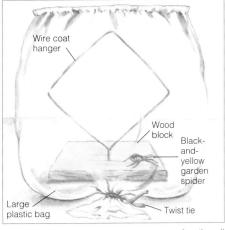

(continued)

Activity Bank

SPINNING WEBS

BEFORE THE ACTIVITY

1. Divide the class into groups of four to six students.
2. Several days before the activity, have students bring in a large, clear plastic bag with a twist tie for each group. The plastic bags should be large enough to fit over the net frame.
3. Gather all materials at least one day prior to the activity.
4. The black-and-yellow garden spider (*Argiope bruennichi*) is a member of a group of spiders called araneids. These spiders are also called orb weavers. They can be purchased from a biological supply company.
5. Have students place a moistened cotton ball on the wood block. This will provide the spider with water.
6. When the activity is completed, have students return the spiders to you. You may keep them in a terrarium for further observations or return them to their natural habitat.

PRE-ACTIVITY DISCUSSION

Have students read the complete What You Must Do (procedure) for the activity.
• **What is the purpose of this activity?** (To observe how spiders spin webs.)
• **What is the purpose of the plastic bag? Why must the plastic bag be clear?** (To keep the spider from escaping; to make it easier to observe the spider without having to remove the plastic bag.)
• **Do you think your spider will build a web?** (Yes.)
• **How many different types of webs do you think the spiders will build?** (Answers will vary. Because the spiders are members of the same species, they will build similar types of webs.)

SAFETY TIPS

Remind students that spiders, like all living organisms, must be treated with care and in a humane manner.

TEACHING STRATEGY

1. Have the groups work together to build their net frames.
2. Students can inflate the plastic bag by holding each of the two corners at the open end of the plastic bag and waving it through the air, instead of blowing into the plastic bag. Then students should quickly put the net frame inside the plastic bag and tie it with a twist tie.
3. After the net frames have been constructed, go from group to group placing the spider inside the plastic bag as one member of each group quickly unties the twist tie.

DISCOVERY STRATEGIES

Discuss how the activity relates to the chapter ideas by asking questions similar to the following.
• **Is a spider an insect?** (No. Insects have six legs and a body with three sections; spiders have eight legs and a body with two sections—relating facts, making comparisons.)
• **Why do spiders build webs?** (To catch their prey—applying concepts.)

WHAT YOU WILL DISCOVER

1. Yes.

2. Webs are usually built every day or night.

3. Answers will vary.

4. Answers will vary. The entire web, or at least the sticky part, is replaced when it loses its stickiness.

5. Answers will vary.

6. Student results should be similar.

GOING FURTHER

Soon after the insects are released into the plastic bag they become entangled in the sticky threads of the spider's web. After an insect is caught in the green-and-yellow garden spider's web, the spider will tie up the insect with silk until the insect is ready to be eaten.

5. Place the net frame in a place where it will remain undisturbed for two days. Record your daily observations in a data table similar to the one shown.

What You Will See

DATA TABLE

Day	Observations
1	
2	

What You Will Discover

1. Did your spider weave a web?

2. What time of day did your spider start to weave its web?

3. Did your spider build just one web?

4. Did your spider rebuild its web or did it build a new web?

5. If your spider built a new web, how often did it do so?

6. Share your results with your classmates. Were their results similiar? Different? Can you explain why?

Going Further

Release flying insects such as fruit flies and regular flies into the plastic bag with your spider. What happens to the insects? How does your spider react?

HOW MANY ARE TOO MANY?

In nature, there is a maximum size of a population that a particular environment can support at any one time. This is known as the carrying capacity. To actually determine the carrying capacity of a population of fruit flies, why not try this activity.

What You Will Need

culture bottle of fruit flies
sheet of unlined white paper
pencil
small paintbrush

What You Must Do 🧪 🐭

1. Carefully examine a culture bottle of fruit flies to identify their basic needs of life.

2. Count the number of fruit flies in the culture bottle. You might find it easier to count the fruit flies when they are inactive. To make the flies inactive, place the bottle of fruit flies in a freezer for one to two minutes. The cold temperature will slow them down.

3. Remove the plug from the culture bottle. Gently shake the fruit flies onto a sheet of unlined white paper and count them. As soon as you see the flies becoming active again, brush them back into the bottle and refrigerate them again, if you need to.

4. Count the number of fruit flies every two days for two weeks. Record this information in a data table similar to the one shown on the next page.

Culture bottle of fruit flies

Culture bottle

Sheet of unlined white paper

Fruit flies

(continued)

C ■ 159

cotton wrapped in cheesecloth. Tilt the bottles to increase the surface and allow the medium to cool. Store the bottles in a refrigerator until the flies are introduced. Just before adding the flies, place a pinch of dry yeast on the surface of the now-solid medium.

4. Gather all materials at least one day prior to the activity.

PRE-ACTIVITY DISCUSSION

Have students read the entire What You Must Do (procedure) for the activity.
• **What is the purpose of this activity?** (To determine the carrying capacity of a population of fruit flies in a culture jar.)
• **What is meant by the term carrying capacity?** (The maximum size of a population that a particular environment can support at any one time.)
• **What materials do the fruit flies need in order to live?** (Oxygen, food, water.)
• **In addition to these materials, what else do fruit flies need?** (Adequate space.)

SAFETY TIPS

1. Remind students that fruit flies, like all living organisms, must be handled with care and in a humane manner.
2. Remind students to be careful when handling breakable materials.

TEACHING STRATEGY

1. Circulate among the groups to make sure that they are recording their fruit-fly numbers every two days.
2. Because the changes in the fruit-fly population are quite dramatic, have students plot these changes on a graph.

DISCOVERY STRATEGIES

Discuss how the activity relates to the chapter ideas by asking questions similar to the following.
• **Are fruit flies insects? Explain.** (Yes. Like all insects, fruit flies have six legs, body made up of three sections—making observations.)
• **How do you think fruit flies got their name?** (They are found around fruit—bananas, apples, and so on—relating facts.)

Activity Bank

HOW MANY ARE TOO MANY?

BEFORE THE ACTIVITY

1. Divide the class into groups of three to six students.
2. Fruit flies (*Drosophila melanogaster*) can be purchased from a biological supply company or you can collect them from fruit and vegetable markets.

3. Several days before the activity, prepare the medium upon which fruit flies live. Dissolve 20 g agar powder in 625 mL water by bringing it to a boil. Stir constantly. To this, add 25 mL white corn syrup and 250 mL banana pulp that was made by mashing a banana with a fork or by putting it in a strainer. Heat the mixture to boiling and quickly pour it into culture bottles to a depth of 1.3 cm. Insert a strip of paper toweling into the medium while it is soft; this will provide more surface for egg laying. Cover the bottles with

WHAT YOU WILL DISCOVER

1. Air, food, water, and space.

2. Food, water, and space are limited because there are only certain amounts of these things contained in the culture bottle. The supply of air is unlimited because the culture bottle is connected to the outside via the porous plug, which allows air to move in and out.

3. Initially, there was an increase in the number of fruit flies by reproduction, causing a decrease in the limited space, water, and food. This decrease causes a decrease in the carrying capacity of the fruit flies' environment.

GOING FURTHER

Students should realize that once a population—be it human or fruit fly—has reached the carrying capacity of its environment, certain factors keep the population from growing any further. These factors include a lack of food, overcrowding, and competition among individuals in the population.

What You Will See

DATA TABLE

Day	Number of Fruit Flies
1	
2	
3	
4	

What You Will Discover

1. What are the four basic needs of life?

2. Which of the four basic needs of life are limited in this activity? Which are unlimited? Explain.

3. What changes did you see in the population of fruit flies?

Going Further

Do you think the information from this activity can be applied to changes in the human population of the world? Explain your answer.

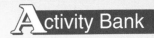

TO FLOAT OR NOT TO FLOAT?

How are most fishes able to swim at various depths in water? Bony fishes can raise or lower themselves in the water by adding gases to or removing them from their swim bladders. A swim bladder is a gas-filled chamber that adjusts the buoyancy of the fish to the water pressure at various depths. Water pressure is the force that water exerts over a certain area due to its weight and motion. To see firsthand how a swim bladder works, try this activity.

Materials

medicine dropper
2-L clear plastic soft drink bottle with cap

Procedure

1. Select one member of your group to act as the Principal Investigator and another to act as the Recorder. The other members of the group will be the Observers.
2. Fill the plastic soft drink bottle to the very top with water.

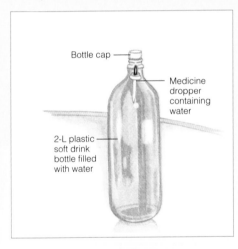

Bottle cap

Medicine dropper containing water

2-L plastic soft drink bottle filled with water

3. Place the medicine dropper in the bottle of water. The water should overflow. You may have to draw in or squeeze out more water from the medicine dropper so that it barely floats.
4. Screw the cap tightly on the plastic bottle. No water or air should leak out when the bottle is squeezed.

Squeezing the Bottle

Releasing the Bottle

5. Squeeze the sides of the bottle. Notice what happens.
6. Release the sides of the bottle. Notice what happens.
7. Have the Investigator and the Recorder reverse roles and repeat steps 5 and 6.

(continued)

A possible hypothesis is: "Some fishes will be able to swim at various depths in water by adding or removing gases from their swim bladders."
• **Why should the cap be screwed on tightly?** (To prevent air or water from leaking out when the bottle is squeezed.)

TEACHING STRATEGY

The key to a successful swim bladder is that the dropper is about 95 percent submerged. If a squeeze of the bottle does not sink the dropper, add more water or siphon a little water up into the dropper.

DISCOVERY STRATEGIES

Discuss how the investigation relates to the chapter ideas by asking questions similar to the following.
• **What does the squeezing and releasing of the bottle do?** (Increases and decreases the pressure inside—predicting, analyzing, relating.)
• **How does increasing the swim bladder's (dropper's) density (mass/volume), affect its buoyancy?** (Its increased density causes it to sink—observing, relating.)
• **What is the relationship between the swim bladder's density and its buoyancy?** (The less its density, the greater its buoyancy. The greater its density, the less its buoyancy—observing, analyzing, relating.)

OBSERVATIONS

1. The medicine dropper sinks.
2. The medicine dropper rises.

ANALYSIS AND CONCLUSIONS

1. The pressure increases.
2. By squeezing the bottle, pressure is being applied. This pressure is felt throughout the liquid. The increased pressure forces water into the medicine dropper.
3. The increased pressure forces water into the medicine dropper, increasing the density of the medicine dropper, so it sinks.
4. When the sides of the bottle are released, the pressure decreases, causing the water to leave the medicine dropper. This action causes the density to decrease and the medicine dropper rises.
5. The fish with less mass. The greater the density (mass divided by volume), the less the buoyancy.

Activity Bank

TO FLOAT OR NOT TO FLOAT?

BEFORE THE ACTIVITY

1. Several days before performing the activity have students bring in a 2-L clear, soft drink bottle with cap. The bottles can be used by different classes and they can also be saved for use in future activites.
2. Divide the class into groups of at least three students.

3. Gather all materials at least one day prior to the activity.

PRE-ACTIVITY DISCUSSION

Have students read the complete procedure.
• **What is the purpose of the activity?** (To observe how a fish's swim bladder works; to observe the effect water pressure has on an object at various depths.)
• **Based on what you already know about the swim bladder of fishes, state a hypothesis.** (Accept any logical hypothesis.

Observations

1. What happens to the medicine dropper when the sides of the bottle are squeezed?
2. What happens to the medicine dropper when the sides of the bottle are released?

Analysis and Conclusions

1. What happens to the pressure of the water when you squeeze the sides of the plastic bottle?
2. Why is some of the water pushed up into the medicine dropper when you squeeze the bottle?
3. Why does the medicine dropper sink when you squeeze the sides of the bottle?
4. Why does the medicine dropper rise when you release the sides of the plastic bottle?
5. Two fishes are very similar in size to each other. However, one has more mass than the other. Which fish has more buoyancy?

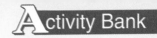

DO OIL AND WATER MIX?

An oil spill is an environmental disaster. Not only can an oil spill be fatal to ocean life but it can also endanger sea birds by damaging their feathers. To find out the effect of water and oil on the feathers of birds, try this activity.

Materials

hand lens
natural bird feather
small container of water
medicine dropper
small container of cooking oil

Procedure

1. Select one member of your group to be the Principal Investigator and another member to act as the Timer. The remaining members of the group will be the Observers.
2. With the hand lens, examine a natural bird feather.
3. Dip the feather into water for 1 minute. Take the feather out of the water and examine it with the hand lens.
4. Dip the feather into the cooking oil for 1 minute. Take it out and examine it with the hand lens.
5. Now dip the oil-covered feather back into the water for 1 minute. Take it out and examine it with the hand lens.

Observations

Describe the appearance of the feather in steps 2 through 5.

Analysis and Conclusions

1. What are the major changes in the feather after being placed in water? After being placed in oil?
2. What happened to the oil-covered feather when it was dipped in water?
3. How does exposure to oil affect normal bird activities?
4. What are some other examples of human-caused pollutants that can have harmful effects on wildlife?

Hand lens

Natural bird feather

Natural bird feather

Hand lens

Activity Bank

DO OIL AND WATER MIX?

BEFORE THE ACTIVITY

1. You can obtain natural bird feathers from a biological supply company.
2. Gather all materials at least one day prior to performing the activity.
3. Divide the class into groups of at least three students.
4. Some students may have an allergy to feathers and should be excused from performing the activity. They also should be assigned to another room when the activity is being done. Prepare an activity in which they are given photographs showing the effects of oil spills on birds.

PRE-ACTIVITY DISCUSSION

Have students read the complete procedure.
• **What is the purpose of this activity?** (To observe the effects water and oil have on a bird feather.)
• **What function do bird feathers serve?** (They provide insulation, create lightweight and streamlined wings for lifting birds.)

TEACHING STRATEGY

1. Circulate about the room as students work to make sure they stay on task.
2. Make sure students do not wipe off the water or oil from the feathers.
3. Students should return feathers to you for proper disposal.

DISCOVERY STRATEGIES

Discuss how the activity relates to the chapter ideas by asking questions similar to the following.
• **What effect does oil have on a bird's feathers?** (Accept all logical answers—observing.)
• **What can be done to prevent oil spills?** (Accept all logical answers—analyzing.)

OBSERVATIONS

When the feather is placed in water, it repels the water. When the feather is placed in oil, it becomes coated with oil. When the oil-coated feather is then dipped in water, the feather loses its ability to repel water.

ANALYSIS AND CONCLUSIONS

1. After being placed in water, the feather remains stiff and keeps its shape. After being placed in oil, the feather becomes slick and coated with oil.
2. The oil-covered feather no longer has the ability to repel water.
3. Because the addition of oil adds more mass to the bird, exposure to oil interferes with a bird's ability to fly. Exposure to oil also reduces the air spaces among the feathers, which provide the bird with insulation, causing a drop in the bird's body temperature. This can cause the bird to die.
4. Pesticides, sewage, chemicals, and so on.

Activity Bank

AN EGGS-AGGERATION

BEFORE THE ACTIVITY

1. Divide students into groups of four students.
2. Several days prior to doing the activity, have each group bring in a hard-boiled egg. Refrigerate the hard-boiled eggs until immediately before the activity.
3. Gather all materials at least one day prior to the activity.
4. The activity requires less than one class period to do and part of a class period the following day. This activity can also be performed by students at home.

PRE-ACTIVITY DISCUSSION

Have students read the complete What You Will Do (procedure) for the activity.
• **What is the purpose of the activity?** (To observe how an egg shell can be removed without cracking it; to see the effect vinegar has on an egg shell.)
• **What do you think will happen to the egg shell?** (Accept all logical answers.)

TEACHING STRATEGY

1. Circulate about the room, making sure students stay on task.
2. Make sure students use tongs to remove the egg from the jar containing vinegar.
3. Collect all materials from students for proper disposal.

DISCOVERY STRATEGIES

Discuss how the activity relates to the chapter ideas by asking questions similar to the following.
• **Why is a bird's egg covered by a shell?** (To provide protection—observing, analyzing.)
• **What is inside the shell?** (White of the egg and the yolk—observing.)

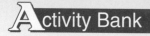
Can you remove the shell from a hard-boiled egg without cracking it? The shell of an egg is made of a substance called calcium carbonate. Calcium carbonate will dissolve in vinegar (acetic acid) to form a solution. Why not try your hand at this activity and see if you can remove the shell without cracking it. You will need a cold hard-boiled egg, a wide-mouthed jar, some vinegar, safety goggles, and tongs.

What You Will Do

1. Place a cold hard-boiled egg in a wide-mouthed jar.
2. Add enough vinegar to cover the egg.
3. Put the jar in a place where it will remain undisturbed overnight.
4. The next morning remove the egg with tongs. Observe the egg.

What You Will See

Describe what happened to the egg.

What You Will Discover

1. What happened to the shell? Did it disappear completely?
2. What do you think happened to the shell?

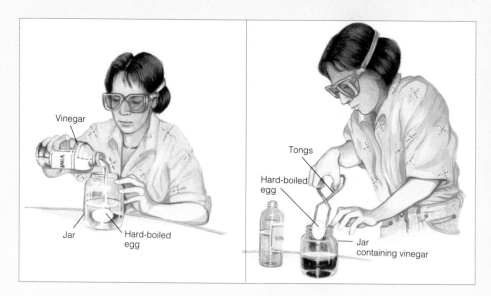

164 ■ C

WHAT YOU WILL SEE

The egg shell disappeared.

WHAT YOU WILL DISCOVER

1. Yes.
2. It dissolved in the vinegar and formed a solution of calcium carbonate and acetic acid.

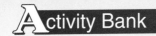

STRICTLY FOR THE BIRDS

Have you ever tried to make a bird's nest? It is not as easy as you may think. Why not try this activity and discover for yourself.

What You Will Do

1. Collect some straw from a field or from along the edges of your yard where the grass has not been mowed.
2. With scissors, cut a handful of straw or grass.
3. Now turn the handful of straw around and around in your hand as shown in the diagram on the left.
4. Try to form the straw into a donut shape using your thumbs to hollow out the middle as shown in the diagram on the right.

What You Will Discover

1. Is your nest very strong?
2. Drop it on a table. Does your nest stay together?
3. Who can make a better bird's nest—you or a bird? Why do you think this is true?
4. What other materials do birds use in making a nest?

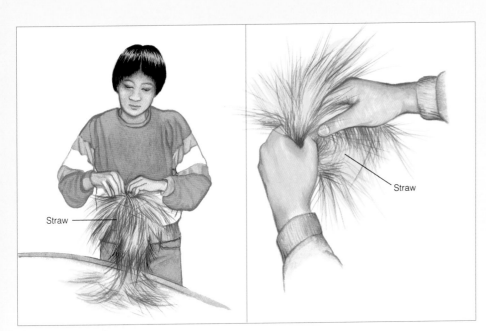

Straw

Straw

TEACHING STRATEGY

1. Circulate about the room, helping students form the straw into a donut shape.
2. You may wish to have materials such as leaves, pine needles, or dead flowers on hand for students. Have them place any or all of these materials inside the nests to make them softer as birds do.

DISCOVERY STRATEGIES

Discuss how the activity relates to chapter ideas by asking questions similar to the following.
• **What structures do birds have that enable them to build a nest?**
(They have beaks and tiny feet—observing, analyzing.)
• **How do birds learn how to build nests?**
(By instinct—analyzing.)

WHAT YOU WILL DISCOVER

1. Answers will vary.
2. Answers will vary.
3. Students should realize that a bird can build a better nest then they can. Not only do birds have more experience in nest building, but they also can find the most suitable materials for their nests.
4. Mud and other plant materials.

Activity Bank

STRICTLY FOR THE BIRDS

BEFORE THE ACTIVITY

1. Divide the class into groups of three students.
2. Gather all materials at least one day prior to the activity.

PRE-ACTIVITY DISCUSSION

Have students read the complete What You Will Do (procedure) for the activity.
• **What is the purpose of this activity?**
(To see how a bird builds its nest, to see if students can build a better nest than a bird can build.)
• **Why did you use straw or grass to build your nest?** (To use the same materials as most birds use when they build their nests.)

Appendix A

The metric system of measurement is used by scientists throughout the world. It is based on units of ten. Each unit is ten times larger or ten times smaller than the next unit. The most commonly used units of the metric system are given below. After you have finished reading about the metric system, try to put it to use. How tall are you in metrics? What is your mass? What is your normal body temperature in degrees Celsius?

Commonly Used Metric Units

Length   The distance from one point to another

meter (m)	A meter is slightly longer than a yard.
	1 meter = 1000 millimeters (mm)
	1 meter = 100 centimeters (cm)
	1000 meters = 1 kilometer (km)

Volume   The amount of space an object takes up

liter (L)	A liter is slightly more than a quart.
	1 liter = 1000 milliliters (mL)

Mass   The amount of matter in an object

gram (g)	A gram has a mass equal to about one paper clip.
	1000 grams = 1 kilogram (kg)

Temperature   The measure of hotness or coldness

degrees	0°C = freezing point of water
Celsius (°C)	100°C = boiling point of water

Metric–English Equivalents

2.54 centimeters (cm) = 1 inch (in.)
1 meter (m) = 39.37 inches (in.)
1 kilometer (km) = 0.62 miles (mi)
1 liter (L) = 1.06 quarts (qt)
250 milliliters (mL) = 1 cup (c)
1 kilogram (kg) = 2.2 pounds (lb)
28.3 grams (g) = 1 ounce (oz)
$°C = 5/9 \times (°F - 32)$

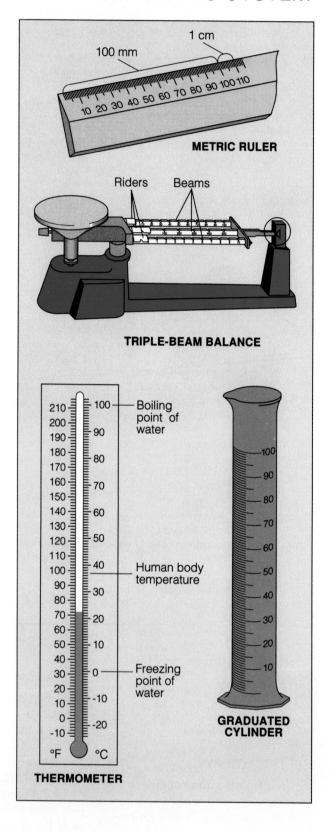

METRIC RULER

TRIPLE-BEAM BALANCE

Riders   Beams

Boiling point of water

Human body temperature

Freezing point of water

°F   °C

THERMOMETER

GRADUATED CYLINDER

Glassware Safety

1. Whenever you see this symbol, you will know that you are working with glassware that can easily be broken. Take particular care to handle such glassware safely. And never use broken or chipped glassware.
2. Never heat glassware that is not thoroughly dry. Never pick up any glassware unless you are sure it is not hot. If it is hot, use heat-resistant gloves.
3. Always clean glassware thoroughly before putting it away.

Fire Safety

1. Whenever you see this symbol, you will know that you are working with fire. Never use any source of fire without wearing safety goggles.
2. Never heat anything—particularly chemicals—unless instructed to do so.
3. Never heat anything in a closed container.
4. Never reach across a flame.
5. Always use a clamp, tongs, or heat-resistant gloves to handle hot objects.
6. Always maintain a clean work area, particularly when using a flame.

Heat Safety

Whenever you see this symbol, you will know that you should put on heat-resistant gloves to avoid burning your hands.

Chemical Safety

1. Whenever you see this symbol, you will know that you are working with chemicals that could be hazardous.
2. Never smell any chemical directly from its container. Always use your hand to waft some of the odors from the top of the container toward your nose—and only when instructed to do so.
3. Never mix chemicals unless instructed to do so.
4. Never touch or taste any chemical unless instructed to do so.
5. Keep all lids closed when chemicals are not in use. Dispose of all chemicals as instructed by your teacher.

6. Immediately rinse with water any chemicals, particularly acids, that get on your skin and clothes. Then notify your teacher.

Eye and Face Safety

1. Whenever you see this symbol, you will know that you are performing an experiment in which you must take precautions to protect your eyes and face by wearing safety goggles.
2. When you are heating a test tube or bottle, always point it away from you and others. Chemicals can splash or boil out of a heated test tube.

Sharp Instrument Safety

1. Whenever you see this symbol, you will know that you are working with a sharp instrument.
2. Always use single-edged razors; double-edged razors are too dangerous.
3. Handle any sharp instrument with extreme care. Never cut any material toward you; always cut away from you.
4. Immediately notify your teacher if your skin is cut.

Electrical Safety

1. Whenever you see this symbol, you will know that you are using electricity in the laboratory.
2. Never use long extension cords to plug in any electrical device. Do not plug too many appliances into one socket or you may overload the socket and cause a fire.
3. Never touch an electrical appliance or outlet with wet hands.

Animal Safety

1. Whenever you see this symbol, you will know that you are working with live animals.
2. Do not cause pain, discomfort, or injury to an animal.
3. Follow your teacher's directions when handling animals. Wash your hands thoroughly after handling animals or their cages.

One of the first things a scientist learns is that working in the laboratory can be an exciting experience. But the laboratory can also be quite dangerous if proper safety rules are not followed at all times. To prepare yourself for a safe year in the laboratory, read over the following safety rules. Then read them a second time. Make sure you understand each rule. If you do not, ask your teacher to explain any rules you are unsure of.

Dress Code

1. Many materials in the laboratory can cause eye injury. To protect yourself from possible injury, wear safety goggles whenever you are working with chemicals, burners, or any substance that might get into your eyes. Never wear contact lenses in the laboratory.

2. Wear a laboratory apron or coat whenever you are working with chemicals or heated substances.

3. Tie back long hair to keep it away from any chemicals, burners and candles, or other laboratory equipment.

4. Remove or tie back any article of clothing or jewelry that can hang down and touch chemicals and flames.

General Safety Rules

5. Read all directions for an experiment several times. Follow the directions exactly as they are written. If you are in doubt about any part of the experiment, ask your teacher for assistance.

6. Never perform activities that are not authorized by your teacher. Obtain permission before "experimenting" on your own.

7. Never handle any equipment unless you have specific permission.

8. Take extreme care not to spill any material in the laboratory. If a spill occurs, immediately ask your teacher about the proper cleanup procedure. Never simply pour chemicals or other substances into the sink or trash container.

9. Never eat in the laboratory.

10. Wash your hands before and after each experiment.

First Aid

11. Immediately report all accidents, no matter how minor, to your teacher.

12. Learn what to do in case of specific accidents, such as getting acid in your eyes or on your skin. (Rinse acids from your body with lots of water.)

13. Become aware of the location of the first-aid kit. But your teacher should administer any required first aid due to injury. Or your teacher may send you to the school nurse or call a physician.

14. Know where and how to report an accident or fire. Find out the location of the fire extinguisher, phone, and fire alarm. Keep a list of important phone numbers—such as the fire department and the school nurse—near the phone. Immediately report any fires to your teacher.

Heating and Fire Safety

15. Again, never use a heat source, such as a candle or burner, without wearing safety goggles.

16. Never heat a chemical you are not instructed to heat. A chemical that is harmless when cool may be dangerous when heated.

17. Maintain a clean work area and keep all materials away from flames.

18. Never reach across a flame.

19. Make sure you know how to light a Bunsen burner. (Your teacher will demonstrate the proper procedure for lighting a burner.) If the flame leaps out of a burner toward you, immediately turn off the gas. Do not touch the burner. It may be hot. And never leave a lighted burner unattended!

20. When heating a test tube or bottle, always point it away from you and others. Chemicals can splash or boil out of a heated test tube.

21. Never heat a liquid in a closed container. The expanding gases produced may blow the container apart, injuring you or others.

22. Before picking up a container that has been heated, first hold the back of your hand near it. If you can feel the heat on the back of your hand, the container may be too hot to handle. Use a clamp or tongs when handling hot containers.

Using Chemicals Safely

23. Never mix chemicals for the "fun of it." You might produce a dangerous, possibly explosive substance.

24. Never touch, taste, or smell a chemical unless you are instructed by your teacher to do so. Many chemicals are poisonous. If you are instructed to note the fumes in an experiment, gently wave your hand over the opening of a container and direct the fumes toward your nose. Do not inhale the fumes directly from the container.

25. Use only those chemicals needed in the activity. Keep all lids closed when a chemical is not being used. Notify your teacher whenever chemicals are spilled.

26. Dispose of all chemicals as instructed by your teacher. To avoid contamination, never return chemicals to their original containers.

27. Be extra careful when working with acids or bases. Pour such chemicals over the sink, not over your workbench.

28. When diluting an acid, pour the acid into water. Never pour water into an acid.

29. Immediately rinse with water any acids that get on your skin or clothing. Then notify your teacher of any acid spill.

Using Glassware Safely

30. Never force glass tubing into a rubber stopper. A turning motion and lubricant will be helpful when inserting glass tubing into rubber stoppers or rubber tubing. Your teacher will demonstrate the proper way to insert glass tubing.

31. Never heat glassware that is not thoroughly dry. Use a wire screen to protect glassware from any flame.

32. Keep in mind that hot glassware will not appear hot. Never pick up glassware without first checking to see if it is hot. See #22.

33. If you are instructed to cut glass tubing, fire-polish the ends immediately to remove sharp edges.

34. Never use broken or chipped glassware. If glassware breaks, notify your teacher and dispose of the glassware in the proper trash container.

35. Never eat or drink from laboratory glassware. Thoroughly clean glassware before putting it away.

Using Sharp Instruments

36. Handle scalpels or razor blades with extreme care. Never cut material toward you; cut away from you.

37. Immediately notify your teacher if you cut your skin when working in the laboratory.

Animal Safety

38. No experiments that will cause pain, discomfort, or harm to mammals, birds, reptiles, fishes, and amphibians should be done in the classroom or at home.

39. Animals should be handled only if necessary. If an animal is excited or frightened, pregnant, feeding, or with its young, special handling is required.

40. Your teacher will instruct you as to how to handle each animal species that may be brought into the classroom.

41. Clean your hands thoroughly after handling animals or the cage containing animals.

End-of-Experiment Rules

42. After an experiment has been completed, clean up your work area and return all equipment to its proper place.

43. Wash your hands after every experiment.

44. Turn off all burners before leaving the laboratory. Check that the gas line leading to the burner is off as well.

Glossary

Pronunciation Key

When difficult names or terms first appear in the text, they are respelled to aid pronunciation. A syllable in SMALL CAPITAL LETTERS receives the most stress. The key below lists the letters used for respelling. It includes examples of words using each sound and shows how the words would be respelled.

Symbol	Example	Respelling
a	hat	(hat)
ay	pay, late	(pay), (layt)
ah	star, hot	(stahr), (haht)
ai	air, dare	(air), (dair)
aw	law, all	(law), (awl)
eh	met	(meht)
ee	bee, eat	(bee), (eet)
er	learn, sir, fur	(lern), (ser), (fer)
ih	fit	(fiht)
igh	mile, sigh	(mighl), (sigh)
oh	no	(noh)
oi	soil, boy	(soil), (boi)
oo	root, rule	(root), (rool)
or	born, door	(born), (dor)
ow	plow, out	(plow), (owt)

Symbol	Example	Respelling
u	put, book	(put), (buk)
uh	fun	(fuhn)
yoo	few, use	(fyoo), (yooz)
ch	chill, reach	(chihl), (reech)
g	go, dig	(goh), (dihg)
j	jet, gently, bridge	(jeht), (JEHNTlee), (brihj)
k	kite, cup	(kight), (kuhp)
ks	mix	(mihks)
kw	quick	(kwihk)
ng	bring	(brihng)
s	say, cent	(say), (sehnt)
sh	she, crash	(shee), (krash)
th	three	(three)
y	yet, onion	(yeht), (UHN yuhn)
z	zip, always	(zihp), (AWL wayz)
zh	treasure	(TREH zher)

asexual reproduction: process by which a single organism produces a new organism

autotroph (AW-toh-trahf): organism that can make its own food

coldblooded: having a body temperature that changes somewhat with the temperature of the surroundings

contour feather: largest and most familiar feather that gives birds their streamlined shape

down: short, fluffy feather that acts as insulation

egg-laying mammal: warmblooded vertebrate with hair or fur that lays eggs; monotreme

external fertilization: process in which a sperm joins with an egg outside the body

exoskeleton: rigid outer covering in most arthropods

feather: important characteristic of birds; helps to insulate the body and is used in flying

gestation (jehs-TAY-shuhn) **period:** time the young of placental mammals spend inside the mother

gill: feathery structure through which water-dwelling animals breathe

heterotroph (HEHT-er-oh-trahf): organism that cannot make its own food

host: organism upon which another organism lives

internal fertilization: process in which a sperm joins with an egg inside the body

invertebrate: animal that has no backbone

kingdom: large general classification group

larva (LAHR-vuh): second stage in metamorphosis, when an egg hatches

metamorphosis (meht-uh-MOR-fuh-sihs): process by which an organism undergoes dramatic changes in body form in its life cycle

migrate: to move to a new environment during the course of a year

molting: process by which an arthropod's exoskeleton is shed and replaced from time to time

nematocyst (NEHM-uh-toh-sihst): stinging structure used by a cnidarian to stun or kill its prey

parasite: organism that grows on or in other living organisms

pheromone (FER-uh-mohn): powerful chemical given off by an insect to attract a mate

placenta (pluh-SEHN-tuh): structure that develops in pregnant female placental mammals through which food, oxygen, and wastes are exchanged between young and mother

placental (pluh-SEHN-tuhl) **mammal:** warmblooded vertebrate with hair or fur that gives birth to young that have remained inside the mother's body until their body systems are able to function independently

pouched mammal: warmblooded vertebrate with hair or fur that gives birth to young that are not well developed; marsupial

pupa (PYOO-puh): third stage in the metamorphosis of an insect

regeneration: ability to regrow lost parts

sexual reproduction: process by which a new organism forms from the joining of a female cell and a male cell

spicule (SPIHK-yool): thin, spiny structure that forms the skeleton of many sponges

swim bladder: gas-filled sac that gives bony fishes buoyancy

territory: area where an individual animal lives

tube foot: suction-cuplike structure connected to the water vascular system of an echinoderm

vertebrate: animal that has a backbone, or vertebral column

warmblooded: having a body temperature that stays constant

water vascular system: fluid-filled internal tubes that carry food and oxygen, remove wastes, and help echinoderms move

Index

African lungfish, C75
Alligators, C101
  characteristics of, C101
  compared to crocodiles, C101
Amphibians, C76–82
  characteristics of, C77–80
  fertilization of eggs, C80
  frogs, C81–82
  looped circulatory system,
    C78–79
  metamorphosis, C79–80
  newts, C82
  respiration of, C77–78
  salamanders, C82
  toads, C81–82
Angler fishes, C68
Animalia, C14
Animals
  cells of, C15
  characteristics of, C14
  invertebrates, C16
  phyla, C16–17
  vertebrates, C15–16
Annelida, C28
Anteaters, C131–132
  spiny anteater, C124
Arachnids
  characteristics of, C44
  members of phylum, C44
  scorpions, C46–47
  spiders, C44–46
  ticks and mites, C47
Archaeopteryx, C103
Armadillos, C131–132
Arrow-poison frogs, C77–78
Arthropoda, C41
Arthropods, C40–52
  arachnids, C44
  centipedes, C44
  characteristics of, C40–42
  crustaceans, C42–43
  evolution of, C41
  exoskeleton, C41
  insects, C48–52
  jointed appendages, C41
  millipedes, C44
  open circulatory system, C41–42
  segmented bodies, C41
  sexual reproduction, C42
  size of phylum, C40
  spiders, C44–47
Asexual reproduction
  cnidarians, C21
  process of, C20
  sponges, C20
Autotrophs, C12–13

Baboons, C136
Bandicoots, C125
Barnacles, C42
Bats, C129
Beavers, C134
Bees, as social insects, C51–52
Behavior
  birds, C107–109
  insects, C51–52
Birds, C103–112
  behavior, C107–109
  birds of prey, C111
  characteristics of, C103–107
  eating habits, C104–105
  eggs, care of, C107
  evolution of, C103
  feathers, C103–104
  flight, C105–106
  flightless birds, C112
  hunting birds, C111
  looped circulatory system, C106
  mating behavior, C108–109,
    C113
  migration, C109
  senses of, C106
  song, meaning of, C107–108
  songbirds, C110–111
  territoriality, C108
  as warmblooded animals, C103
  waterfowl, C111
  young, care of, C107
Birds of prey, C111
Bivalves, C33
  characteristics of, C32
Bony fishes, C73–75
  adaptations of, C74–75
  characteristics of, C73–75
  fins, C73
  swim bladders, C73–74
Book lungs, spiders, C45–46
Bowerbirds, C108–109
Bubonic plague, C134

Camels, C133
Camouflage, insects, C52
Canaries, C110
Canine teeth, C130–131
Cardinals, C110
Carnivores, C131
Cartilage, C70, C72
Cartilaginous fishes, C71–73
  characteristics of, C71–73
  rays, C72–73
  sharks, C71–72
  skates, C72–73

Cassowary, C112
Cells, animal, C15
Centipedes, C44
  characteristics of, C44
Cephalopods, C33
Chambered nautilus, C33
Chameleons, C96, C97
Chimaeras, C71
Chimpanzee, C136
Chipmunks, C134
Chordata, C64
  characteristics of, C64
Chrysalis, C50
Cicada, C51
Clams, C32
Closed circulatory system
  earthworms, C28–29
  fishes, C68
Clownfish, C24
Cnidarians, C20–24
  asexual reproduction, C21
  characteristics of, C20–21
  corals, C22–23
  hydras, C22
  jellyfishes, C24
  nematocysts, C21
  phylum of, C20
  sea anemones, C24
  sexual reproduction, C21
Cocoon, C50
Coelacanths, C73, C77
Coldblooded animals, C66, C91
Compound eyes, insects, C49
Contour feathers, C104
Cook, James, C126
Copperheads, C99
Coral, C22–23
  and bone replacement surgery,
    C23
  characteristics of, C22–23
Coral snakes, C99
Crabs, C42, C43
  regeneration, C43
Crocodiles, C101
  characteristics of, C101
  compared to alligators, C101
Crows, C110
Crustaceans, C42–43
  characteristics of, C42–43
  gills, C43
  regeneration, C43

Deep-sea fishes, C74
Deer, C133

Defense mechanisms, insects, C52
Dinosaurs, C89
Dolphins, C135
Down, feathers, C104
Dragonfly, flight of, C57
Dragons, Komodo, C95
Duckbill platypus, C123–124
Ducks, C111
Dugongs, C135

Eagles, C111
Earthworms, C16, C28–29
Echinodermata, C54
Echinoderms, C54–56
   characteristics of, C54
   feather stars, C56
   sand dollars, C56
   sea cucumbers, C56
   sea lilies, C56
   sea urchins, C56
   starfishes, C55
Ectotherms, C66
Egg-laying mammals, C123–124
   duckbill platypus, C123–124
   spiny anteater, C124
Electric eel, C73
Elephant, C132–133
Emu, C112
Endoskeleton, nature of, C64
Endotherms, C66
Exoskeleton
   arthropods, C41
   centipedes and millipedes, C44
   crustaceans, C42
   insects, shedding of, C49
   molting, C41
External fertilization, C69

Falcon, C111
Feathers
   contour feathers, C104
   down, C104
Feather stars, C56
Fertilization of eggs, C69
   amphibians, C80
   external, C69
   fishes, C69
   internal, C69, C80, C93, C122–123
   mammals, C122–123
   reptiles, C93–94
Filter feeders, bivalves, C32
Fins, C67, C73

Fireflies, C39, C51
Fishes, C67–75
   bony, C73–75
   cartilaginous, C71–73
   characteristics of, C67–68
   closed circulatory system, C68
   evolution of, C67
   fertilization of eggs, C69
   jawless, C70–71
   mating behavior, C70
   schools, C73
   sexual reproduction, C69
Five kingdoms
   Animalia, C14
   Fungi, C13–14
   Monera, C12–13
   Plantae, C14
   Protista, C13
Flatworms, C26–27
   characteristics of, C26
   parasites, C26–27
   phylum of, C26
   planarian, C26
   regeneration, C26
   tapeworms, C26–27
Flesh-eating mammals, C130–131
Flight
   birds, C105–106
   flying mammals, C129
   insects, C49, C57
Flightless birds, C112
Flounder, C74
Frogs, C81–82
   characteristics of, C81–82
   compared to toads, C81–82
Fungi
   characteristics of, C13–14
   compared to plants, C14
   kingdom, C13–14

Gastropods, C31
Gestation period, C128
Gila monster, C96
Gills
   amphibians, C77
   chordates, C64
   crustaceans, C43
   fishes, C67
Giraffe, C133
Gliders, C125
Gnawing mammals, C134
Goats, C133
Gorilla, C136
Great Barrier Reef, C22

Hagfishes, C71
Hares, C134
   compared to rabbits, C135
Hawks, C111
Hearts, and closed circulatory system, C29
Hedgehogs, C129
Herbivores, C126
Hermit crab, C42
Heterotrophs, C13
Hibernation, frogs and toads, C81
Hippopotamus, C133
Hoofed mammals, C133
Hookworm, C27
Horses, C133
Host, and parasites, C27
Humans, C136
Hunting birds, C111
Hydras, C22
   characteristics of, C22

Ink, of octopus and squid, C33
Insect-eating mammals, C128–129
Insects, C48–52
   behavior of, C51–52
   body structure of, C48–49
   compound eyes, C49
   defense mechanisms, C52
   flight, C49, C57
   metamorphosis, C49–51
   sexual reproduction, C49
Internal fertilization, C69, C80, C93, C122–123
Invertebrates, C16
   arthropods, C40–52
   centipedes, C44
   characteristics of, C16
   cnidarians, C20–24
   corals, C22–23
   crustaceans, C42–43
   echinoderms, C54–56
   hydras, C22
   insects, C48–52
   jellyfishes, C24
   millipedes, C44
   mollusks, C30–33
   sea anemones, C24
   sea cucumbers, C56
   sea lilies, C56
   sea urchins, C56
   spiders, C44–47
   sponges, C18–20
   starfishes, C55
   worms, C25–29

Jawless fishes, C70–71
   characteristics of, C70–71
Jellyfishes, C16, C24
   characteristics of, C24
Jointed appendages, arthropods, C41

Kangaroo, C126
Kingdoms, C12
   See also Five kingdoms.
Koala, C125–126
Komondo dragons, C95

Lampreys, C71
Lantern fishes, C74
Larva, C49
Lizards, C94–97
   characteristics of, C95–97
Lobe-finned fishes, C73
Lobsters, C42
Looped circulatory system, C78–79
   birds, C106
   reptiles, C92
Lyme disease, C47

Mammals, C119–136
   characteristics of, C120–123
   circulatory system of, C121
   egg-laying, C123–124
   evolution of, C120
   excretory system of, C121
   milk production, C120
   nervous system of, C121
   placental, C127–136
   pouched, C125–127
   reproductive methods, C123
   respiration of, C121
   senses, C122
   as warmblooded animals, C120
Mammary glands, C120
Manatees, C135
Marsupials. See Pouched mammals.
Metamorphosis
   amphibians, C79–80
   complete metamorphosis, C49–50
   incomplete metamorphosis, C50–51
   insects, C49–51
   larva, C49
   pupa, C50

Mice, C134
Migration, birds, C109
Mildew, as fungi, C13
Milk production, mammals, C120
Millipedes, characteristics of, C44
Mites, C47
Mockingbirds, C110
Molds, as fungi, C13
Mole, star-nosed, C128–129
Mollusca, C30
Mollusks, C30–33
  bivalves, C33
  cephalopods, C33
  characteristics of, C30–31
  gastropods, C31
  phylum of, C30
  slugs, C31
  snails, C31
  tentacled mollusks, C33
  two-shelled mollusks, C32
Molting, C50–51
  exoskeleton, C41
Monera, C12–13
Monerans
  autotrophs, C12–13
  characteristics of, C12–13
  heterotrophs, C13
Monkeys, C136
Monotremes. *See* Egg-laying mammals.
Mosquitoes, C16
Mucus, C28, C71
Mud puppy, C82
Mudskippers, C75
Mushrooms, as fungi, C13
Mussels, C32

Nematocysts, C24
  cnidarians, C21
Nematoda, C27
Nerve cord, of chordates, C64
Nerve net, C21
Nervous system
  amphibians, C79
  fishes, C68
  mammals, C120–121
Nests, of birds, C109
Newts, C82
Nightingales, C110
Notochord, C64
Nudibranchs, C31

Octopuses, C33
Open circulatory system,

arthropods, C41–42
Opossums, C127
Orangutan, C136
Ostrich, C112
Owls, C111
Oysters, C32

Parasites
  flatworms, C26–27
  and host, C27
Pearl, formation of, C32
Penguins, C112
Perches, C73
Pheromones, C51
Phyla, C16–17
Pigs, C133
Pikas, C134, C135
Placental mammals, C123, C127–136
  characteristics of, C127–128
  flesh-eating, C130–131
  flying, C129
  gnawing, C134
  hoofed, C133
  insect-eating, C128–129
  primates, C136
  rodentlike, C134–135
  toothless, C131–132
  trunk-nosed, C132–133
  water-dwelling, C135
Planarian, C26
Plantae, C14
Plants, C14
Platyhelminthes, C26
Porcupines, C134
Porifera, C18
Porpoises, C135
Pouched mammals, C123, C125–127
  characteristics of, C125
  as herbivores, C126
  kangaroo, C126
  koala, C125–126
  opossums, C127
Primates, C136
Protista, C13
Protists, characteristics of, C13
Pupa, C50
Pygmy shrew, C129

Rabbits, C134
  compared to hares, C135
Rats, C134
Rattlesnakes, C99
Ravens, C110
Ray-finned fishes, C73
Rays, C72–73

Regeneration
  crabs, C43
  crustaceans, C43
  flatworms, C26
  starfishes, C55
Remora, C73
Reptiles, C90–101
  alligators, C101
  characteristics of, C90–94
  as coldblooded animals, C91
  crocodiles, C101
  development of embryo, C92–93
  evolution of, C91
  fertilization of egg, C93–94
  lizards, C94–97
  looped circulatory system, C92
  skin of, C91
  snakes, C97–98
  turtles, C99–100
Rhea, C112
Rhinoceroses, C133
Robins, C110
Rocky Mountain spotted fever, C47
Rodentlike mammals, C134–135
Rodents, C134
Roundworms, C27–28
  characteristics of, C27–28
  hookworm, C27
  phylum of, C27
  trichinosis, C27

Salamanders, C82
Sand dollars, C56
Sawfishes, C71
Scales, fishes, C67
Scallops, C32
Scorpions, C46–47
Sea anemones, C24
Sea butterflies, C31
Sea cucumbers, C56
Sea horses, C73
Sea lilies, C56
Seals, C131
Sea otter, C119
Sea turtles, C100
Sea urchins, C56
Segmented bodies, arthropods, C41
Segmented worms, C28–29, C41
  characteristics of, C28–29
  earthworm, C28–29
  phylum of, C28

Setae, C28
Sexual reproduction
  arthropods, C42
  cnidarians, C21
  fishes, C69
  insects, C49
  and pheromones, C51
  process of, C19
  sponges, C19–20
Sharks, C68, C71–72
Shrew, pygmy, C129
Shrimp, C42
Skates, C72–73
Sloths, C132
Slugs, C31
Snails, C31
Snakes, C97–98
  characteristics of, C97–98
  detection of food, C99
  poisonous snakes, C99
Social insects, C51–52
Songbirds, C110–111
Sparrows, C110
Spicules, sponges, C19
Spiders, C44–47
  book lungs, C45–46
  characteristics of, C44–46
  habitats of, C46
  prey of, C44–45
  webs, C45
Spiny anteater, C124
Sponges, C18–20
  asexual reproduction, C20
  characteristics of, C18–19
  phylum of, C18
  sexual reproduction, C19–20
  spicules, C19
Squids, C33
Squirrels, C134
Starfishes, C55
  characteristics of, C55
  and clams, C55
  regeneration, C55
Star-nosed mole, C128–129
Swans, C111
Swim bladders, C73–74
Swordfishes, C68

Tadpole, metamorphosis, C79–80
Talons, C111
Tapeworms, C26–27
Tapirs, C133
Tentacled mollusks, C33
  characteristics of, C33
Territoriality, birds, C108

Ticks, C47
Toad fishes, C68
Toads
  characteristics of,
    C81–82
  compared to frogs,
    C81–82
  poison of, C82
  and wart myth, C83
Toothless mammals,
  C131–132
Trichinella, C27
Trichinosis, C27
Trunk-nosed mammals,
  C132–133
Tube feet, echinoderms,
  C54

Turtles, C99–100
  characteristics of,
    C99–100
  sea turtles, C100

Ultrasonic sound, C137
Urea, C121

Vertebral column, of
  vertebrates, C64
Vertebrates, C15–16
  amphibians, C76–82
  birds, C103–112
  characteristics of, C16,
    C64

coldblooded animals,
  C66
fishes, C67–75
mammals, C119–136
phylum of, C64, C66
reptiles, C90–101
warmblooded animals,
  C66
Walruses, C130–131
Warblers, C110
Warmblooded animals,
  C66
  birds, C103
  mammals, C120
Wasps, C52
Water-dwelling mammals,
  C135

Waterfowl, C111
Water moccasins, C99
Water vascular system
  echinoderms, C54
Weaverbirds, C109
Webs, spiders, C45
Whales, C135
Wombats, C125
Worms, C25–29
  flatworms, C26–27
  phyla of, C25
  roundworms, C27–28
  segmented worms,
    C28–29

Zebras, C133

Cover Background: Ken Karp
Photo Research: Omni-Photo Communications, Inc.
Contributing Artists: Holly Jones/Cornell & McCarthy, Art Representatives; Kim Mulkey; Ray Smith; Warren Budd Assoc., Ltd.; Fran Milner; Don Martinetti; Mark Schuller
Photographs: 4 top: Charles Seaborn/Odyssey Productions; bottom: R. Andrew Odum/Peter Arnold, Inc.; **5** left: John Gerlach/Tony Stone Worldwide/Chicago Ltd.; right: Brian Parker/Tom Stack & Associates; **6** top: Lefever/Grushow/Grant Heilman; center: Index Stock; bottom: Rex Joseph; **8** left: Fred Bavendam/Peter Arnold, Inc.; right: Hans Pfletschinger/Peter Arnold, Inc.; **9** left: Gary Milburn/Tom Stack & Associates; right: J. Carmichael, Jr./Image Bank; **10** and **11** Charles Seaborn/Odyssey Productions; **12** top: David Muench/David Muench Photography Inc.; bottom: Robert Frerck/Odyssey Productions; **13** top left and bottom: David M. Phillips/Visuals Unlimited; top right: Manfred Kage/Peter Arnold, Inc.; **14** top: Rod Planck/Tom Stack & Associates; bottom left: Milton Rand/Tom Stack & Associates; bottom center: Bradley Smith/Animals Animals/Earth Scenes; bottom right: Zig Leszczynski/Animals Animals/Earth Scenes; **16** top left: St. Meyers/Okapia/Photo Researchers, Inc.; top right: Wm. Curtsinger/Photo Researchers, Inc.; bottom left: Gay Bumgarner/Tony Stone Worldwide/Chicago Ltd.; bottom right: C. A. Morgan/Peter Arnold, Inc.; **18** top right: Fred Bavendam/Peter Arnold, Inc.; bottom left: Mike Schick/Animals Animals/Earth Scenes; bottom center: Carl Roessler/Animals Animals/Earth Scenes; bottom right: Charles Seaborn/Odyssey Productions; **19** Doug Perrine; **20** G. I. Bernard/Oxford Scientific Films/Animals Animals/Earth Scenes; **22** top: G. I. Bernard/Oxford Scientific Films/Animals Animals/Earth Scenes; center left: Allan Power/National Audubon Society/Photo Researchers, Inc.; center right: Denise Tackett/Tom Stack & Associates; bottom left: Andrew Martinez/Photo Researchers, Inc.; bottom right: Charles Seaborn/Odyssey Productions; **23** Charles Seaborn/Odyssey Productions; **24** top: DPI; center: Gregory Dimijian/Photo Researchers, Inc.; bottom: Charles Seaborn/ Odyssey Productions; **25** left: Mike Neumann/Photo Researchers, Inc.; right: G. I. Bernard/Oxford Scientific Films/Animals Animals/Earth Scenes; **26** left: T. E. Adams/Visuals Unlimited; right: Michael Abbey/Photo Researchers, Inc.; **27** top: CNRI/Science Photo Library/Photo Researchers, Inc.; center: R. Calentine/Visuals Unlimited; bottom: Edward Gray/Science Photo Library/Photo Researchers, Inc.; **28** top: CNRI/Science Photo Library/Photo Researchers, Inc.; bottom: Raymond A. Mendez/Animals Animals/Earth Scenes; **29** Breck P. Kent/Animals Animals/Earth Scenes; **30** David M. Dennis/Tom Stack & Associates; **31** top: J. H. Robinson/Animals Animals/Earth Scenes; center: Dr. Paul A. Zahl/Photo Researchers, Inc.; bottom: Dave Fleetham/Tom Stack & Associates; **32** top left: Ed Reschke/Peter Arnold, Inc.; top right: Larry Madin/Planet Earth Pictures; bottom left and right: Dr. Paul A. Zahl/Photo Researchers, Inc.; **33** top: Douglas Faulkner/Photo Researchers, Inc.; center: Tom McHugh/Steinhart Aquarium/Photo Researchers, Inc.; bottom: Fred Bavendam/ Peter Arnold, Inc.; **37** Paulette Brunner/Tom Stack & Associates; **38** and **39** Dr. Ivan Polunin/Bruce Coleman, Inc.; **40** top left: Tom McHugh/Steinhart Aquarium/Photo Researchers, Inc.; top right: Karl Weidmann/Photo Researchers, Inc.; bottom left: Belinda Wright/DRK Photo; bottom right: Don & Pat Valenti/DRK Photo; **41** top: Oxford Scientific Films/Animals Animals/Earth Scenes; center: Don and Esther Phillips/Tom Stack & Associates; bottom left: Brian Parker/Tom Stack & Associates; bottom right: Catherine Ellis/Science Photo Library/Photo Researchers, Inc.; **42** top left: Stephen Dalton/Oxford Scientific Films/Animals Animals/Earth Scenes; top right: Ann Hagen Griffiths/Omni-Photo Communications, Inc.; bottom left: Stephen Dalton/Animals Animals/Earth Scenes; bottom right: Robert Dunne/Photo Researchers, Inc.; **43** top: Chesher/Photo Researchers, Inc.; center: Jett Britnell/DRK Photo; bottom: Tom McHugh/Steinhart Aquarium/Photo Researchers, Inc.; **44** top: Darlyne Murawski/Tony Stone Worldwide/Chicago Ltd.; bottom: David M. Dennis/Tom Stack & Associates; **45** top: George D. Dodge & Dale R. Thompson/Tom Stack & Associates; bottom: Rod Planck/Tom Stack & Associates; **46** top left: Stanley Breeden/DRK Photo; top right: Tom McHugh/Photo Researchers, Inc.; bottom left: Dr. Paul A. Zahl/Photo Researchers, Inc.; bottom right: Raymond A. Mendez/Animals Animals/Earth Scenes; **47** left: Cath Wadforth/Science Photo Library/Photo Researchers, Inc.; center: Belinda Wright/DRK Photo; right: Tom McHugh/Steinhart Aquarium/Photo Researchers, Inc.; **48** Partridge Films Limited/Oxford Scientific Films/Animals Animals/Earth Scenes; **49** top: Kjell B. Sandved; center: John Gerlach/Tony Stone Worldwide/Chicago Ltd.; bottom: Don & Pat Valenti/DRK Photo; **50** left: Breck P. Kent/Animals Animals/Earth Scenes; center left and right: Raymond A. Mendez/Animals Animals/Earth Scenes; right: Alvin E. Staffan/Photo Researchers, Inc.; **51** top: Dwight Kuhn/DRK Photo; bottom left: Stanley Breeden/DRK Photo; bottom center: Ted Clutter/Photo Researchers, Inc.; bottom right: Irvin L. Oakes/Photo Researchers, Inc.; **52** top left: Michael Fogden/DRK Photo; top right and bottom left: James L. Castner; bottom right: Thomas Eisner and Daniel Aneshansley, Cornell University; **54** Thomas Kitchin/Tom Stack & Associates; **55** top: Twomey/Photo Researchers, Inc.; bottom: Fred Bavendam/Peter Arnold, Inc.; **56** top: Charles Seaborn/Odyssey Productions; center top: Andrew J. Martinez/Photo Researchers, Inc.; center bottom: Carl Roessler/Animals Animals/Earth Scenes; bottom: Brian Parker/Tom Stack & Associates; **57** Stephen Dalton/Photo Researchers, Inc.; **58** and **61** J. H. Robinson/Photo Researchers, Inc.; **62** and **63** Dr. Paul A. Zahl/Photo Researchers, Inc.; **64** top: Kevin Schafer/Martha Hill/Tom Stack & Associates; bottom: Merlin Tuttle/Bat Conservation International/Photo Researchers, Inc.; **66** left: Dan Guravich/Photo Researchers, Inc.; right: DPI; **68** top: Jeffrey L. Rotman/Peter Arnold, Inc.; center: Tom McHugh/Steinhart Aquarium/Photo Researchers, Inc.; bottom: Norbert Wu/Peter Arnold, Inc.; **69** left: Ken Lucas/Planet Earth Pictures; right: Zig Leszczynski/Animals Animals/Earth Scenes; **70** left: Marty Snyderman; top right: Tom McHugh/Photo Researchers, Inc.; bottom right: Robert Maier/Animals Animals/Earth Scenes; **71** top: Breck P. Kent; bottom: Ken Lucas/Planet Earth Pictures; **72** left: Marty Snyderman; top right: Tom McHugh/Sea World/Photo Researchers, Inc.; bottom right: Charles Seaborn/Odyssey Productions; **73** top: Lynn Funkhouser/Peter Arnold, Inc.; center: Carl Roessler/Planet Earth Pictures; bottom: Jeff Rotman/Peter Arnold, Inc.; **74** top left: Zig Leszczynski/Animals Animals/Earth Scenes; top right and bottom left: Marty Snyderman; bottom right: Dr. J. Metzner/Peter Arnold, Inc.; **75** left: Tom McHugh/Steinhart Aquarium/Photo Researchers, Inc.; top right: Fred McConnaughey/Photo Researchers, Inc.; bottom right: Tom McHugh/Photo Researchers, Inc.; **76** left: Dr. Paul A. Zahl/Photo Researchers, Inc.; right: Tom McHugh/Steinhart Aquarium/Photo Researchers, Inc.; **77** left: R. Andrew Odum/Peter Arnold, Inc.; top right: Phil A. Dotson/Photo Researchers, Inc.; bottom right: Jany Sauvanet/Photo Researchers, Inc.; **80** top and bottom right: David M. Dennis/Tom Stack & Associates; center left: Hans Pfletschinger/Peter Arnold, Inc.; center right: Nuridsany et Perennou/Photo Researchers, Inc.; bottom left: Oxford Scientific Films/Animals Animals/Earth Scenes; **81** top left: Stephen Dalton/Animals Animals/Earth Scenes; right: Suzanne L. Collins and Joseph T. Collins/Photo Researchers, Inc.; bottom left: Zig Leszczynski/Animals Animals/Earth Scenes; **82** left: E. R. Degginger/Animals Animals/Earth Scenes; right and **83** Breck P. Kent/Animals Animals/Earth Scenes; **87** Tom McHugh/Steinhart Aquarium/Photo Researchers, Inc.; **90** top left: Michael Fogden/Animals Animals/Earth Scenes; top right: Cris Crowley/Tom Stack & Associates; bottom left: Tom Ulrich/Tony Stone Worldwide/Chicago Ltd.; bottom right: Stan Osolinski/Tony Stone Worldwide/Chicago Ltd.; **92** left: Joe B. Blossom/Photo Researchers, Inc.; center: Nancy Adams/Tom Stack & Associates; right and **93** Zig Leszczynski/Animals Animals/Earth Scenes; **94** top: E. R. Degginger/Animals Animals/Earth Scenes; bottom: Leonard Lee Rue III/Tony Stone Worldwide/Chicago Ltd.; **95** left: L.L.T. Rhodes/Animals Animals/Earth Scenes; top right: J. Zerschling/Photo Researchers, Inc.; bottom right: Zig Leszczynski/Animals Animals/Earth Scenes; **96** left: Zig Lesczynski/Animals Animals/Earth Scenes; top right: W. H. Muller/Peter Arnold, Inc.; bottom right: Stephen Dalton/Animals Animals/Earth Scenes; **97** top left: Stephen Dalton/Animals Animals/Earth Scenes; top right: Bob McKeever/Tom Stack & Associates; bottom: Jim Brandenburg/Minden Pictures, Inc.; **98** top left: Mike Severns/Tom Stack & Associates; right and bottom left: Tom McHugh/Photo Researchers, Inc.; **99** top left: John Mitchell/Photo Researchers, Inc.; top right: John Cancalosi/Tom Stack & Associates; bottom left: Tom McHugh/Photo Researchers, Inc.; bottom right: Jany Sauvanet/Photo Researchers, Inc.; **100** top left: Nicholas Parfitt/Tony Stone Worldwide/Chicago Ltd.; top right: Jany Sauvanet/Photo Researchers, Inc.; center left: Jerry L. Ferrara/Photo Researchers, Inc.; center right: Tom McHugh/Steinhart Aquarium/Photo Researchers, Inc.; left top: Brian Parker/Tom Stack & Associates; left bottom: Fred Whitehead/Animals Animals/Earth Scenes; **101** top: Jonathon Blair/Woodfin Camp & Associates; center: Nancy Adams/Tom Stack & Associates; bottom left: Joel Greenstein/Omni-Photo Communications, Inc.; bottom right: Tom McHugh/Photo Researchers, Inc.; **102** David G. Barker/Tom Stack & Associates; **103** Breck P. Kent/Animals Animals/Earth Scenes; **104** left: E. R. Degginger/Animals Animals/Earth Scenes; center: Stouffer Productions/Animals Animals/Earth Scenes; right: Fritz Prenzel/Tony Stone Worldwide/Chicago Ltd.; **105** top: Kevin Schafer/Martha Hill/Tom Stack & Associates; inset, top: Jerome Wexler/Photo Researchers, Inc.; inset, bottom: Tom and Pat Leeson/Photo Researchers, Inc.; center left: Chip and Jill Isenhart/Tom Stack & Associates; center right: Bud Lehnhausen/Photo Researchers, Inc.; bottom: Calvin Larsen/Photo Researchers, Inc.; **106** top: David C. Fritts/Animals Animals/Earth Scenes; bottom: H. A. Thornhill, National Audubon Society/Photo Researchers, Inc.; **107** top: Anna Zuckerman/Tom Stack & Associates; center: Dan Guravich/Photo Researchers, Inc.; bottom left: Scott Camazine/Photo Researchers, Inc.; bottom right: Charles Palek/Animals Animals/Earth Scenes; **108** top left: Richard Kolar/Animals Animals/Earth Scenes; top right: John Garrett/Tony Stone Worldwide/Chicago Ltd.; center right: Bruce Davidson/Animals Animals/Earth Scenes; bottom left: Ron Austing/Photo Researchers, Inc.; **109** left: John Gerlach/Animals Animals/Earth Scenes; center: Jeffrey L. Rotman; right: Hans and Judy Beste/Animals Animals/Earth Scenes; **110** top left: Arthur Gloor/Animals Animals/Earth Scenes; bottom left: Hans Reinhard/Tony Stone Worldwide/Chicago Ltd.; bottom right: Alan G. Nelson/Animals Animals/Earth Scenes; **111** left: Jany Sauvanet/Photo Researchers, Inc.; top right: Harold E. Wilson/Animals Animals/Earth Scenes; bottom right: John Chellman/Animals Animals/Earth Scenes; **112** top left: Manfred Danneger/Tony Stone Worldwide/Chicago Ltd.; top right: Jean-Luc Cherville/Tony Stone Worldwide/Chicago Ltd.; center: Tom McHugh/Photo Researchers, Inc.; bottom left: Frans Lanting/Minden Pictures, Inc.; bottom right: R. D. Estes/Photo Researchers, Inc.; **113** left: Frans Lanting/Minden Pictures, Inc.; right: Jack Vartoogian; **117** Tom Evans/Photo Researchers, Inc.; **118** and **119** Jeff Foott Productions; **120** top: Warren and Genny Garst/Tom Stack & Associates; bottom left: Jeff Lepore/Photo Researchers, Inc.; bottom right: Stephen J. Krasemann/DRK Photo; **121** left: Phil Dotson/DPI; right: Dominique Braud/Tom Stack & Associates; **122** Jerry L. Ferrara/Photo Researchers, Inc.; **124** top: Tom McHugh/Taronga Zoo, Sydney/Photo Researchers, Inc.; bottom: Tom McHugh/Photo Researchers, Inc.; **125** top and bottom left: Dave Watts/Tom Stack & Associates; bottom right: John Cancalosi/Tom Stack & Associates; **126** top: Tony Stone Worldwide/Chicago Ltd.; bottom left: Brian Parker/Tom Stack & Associates; bottom right: E. R. Degginger/Animals Animals/Earth Scenes; **128** top left: Joe McDonald/Animals Animals/Earth Scenes; top right: Rod Planck/Tom Stack & Associates; bottom left: C. O. Harris/Photo Researchers, Inc.; bottom right: Leonard Lee Rue III/Tom Stack & Associates; **129** top: Stephen Dalton/Animals Animals/Earth Scenes; bottom: Oxford Scientific Films/Animals Animals/Earth Scenes; **130** top left: Gerard Lacz/Peter Arnold, Inc.; top right: Pat and Tom Leeson/Photo Researchers, Inc.; bottom left: Tim Davis/Photo Researchers, Inc.; bottom right: Leonard Lee Rue III/Photo Researchers, Inc.; **131** Ira Block/Image Bank; **132** top left: John Cancalosi/Peter Arnold, Inc.; top right: Frederick J. Dodd/Peter Arnold, Inc.; center: Warren Garst/Tom Stack & Associates; bottom: Nicholas Parfitt/Tony Stone Worldwide/Chicago Ltd.; **133** left: Thomas Kitchin/Tom Stack & Associates; right: Frans Lanting/Minden Pictures, Inc.; **134** top: Karl Maslowski/Photo Researchers, Inc.; center: Leonard Lee Rue III/DPI; bottom: John Cancalosi/Tom Stack & Associates; **135** top: Thomas Kitchin/Tom Stack & Associates; center: Robert J. Herko/Image Bank; bottom left: D. Holden Bailey/Tom Stack & Associates; bottom right: James D. Watt/Animals Animals/Earth Scenes; **136** top: Michael Dick/Animals Animals/Earth Scenes; center: Charlie Palek/Animals Animals/Earth Scenes; bottom: Evelyn Gallardo/Peter Arnold, Inc.; **137** left: Tony Stone Worldwide/Chicago Ltd.; right: Stephen Dalton/Oxford Scientific Films/Animals Animals/Earth Scenes; **141** C. C. Lockwood/Animals Animals/Earth Scenes; **142** Duncan Willets/Camerapix; **143** top: Frans Lanting/Minden Pictures, Inc.; bottom: Belinda Wright/DRK Photo; **144** and **145** left: Roger Grace/Greenpeace; right: National Marine Fisheries Services/Earthtrust; bottom: Leonard Lessin/Peter Arnold, Inc.; **146** top: Wm. Curtsinger/Photo Researchers, Inc.; bottom: Leonard Lessin/Peter Arnold, Inc.; **149** Alex Borodulin/Leo De Wys, Inc.; **150** Tom McHugh/Steinhart Aquarium/Photo Researchers, Inc.; **170** Wm. Curtsinger/Photo Researchers, Inc.; **172** J. Zerschling/Photo Researchers, Inc.